T0241901

Research and Perspectives in Endocrine Interactions

More information about this series at http://www.springer.com/series/5241

Donald Pfaff • Yves Christen
Editors

Stem Cells in Neuroendocrinology

 Springer

OPEN

Editors
Donald Pfaff
Department of Neurobiology & Behavior
The Rockefeller University
New York, New York
USA

Yves Christen
Fondation Ipsen
Boulogne Billancourt
France

ISSN 1861-2253   ISSN 1863-0685   (electronic)
Research and Perspectives in Endocrine Interactions
ISBN 978-3-319-82394-2   ISBN 978-3-319-41603-8   (eBook)
DOI 10.1007/978-3-319-41603-8

Introduction

Techniques for manipulating neural systems in general and neuroendocrine systems in particular have matured greatly compared to the era in which nerve cell destruction and electrical stimulation provided our main tools. In theory, nerve cell groups connected with hormonal systems should offer strategic advantages to the stem cell biologist because of the wealth of chemically understood regulatory steps to exploit. While the current volume cannot provide a comprehensive review of the quickly evolving applications of stem cell biology, it does provide a first view of some of the early successes and new possibilities.

For example, the striking successes of Lorenz Studer with dopamine-expressing neurons may not only prove to be of surpassing importance for Parkinson's disease but may also shed light on dopaminergic neuron participation in basic processes of behavioral reward. Inna Tabansky, in addition, portrays how neuroendocrine neurons derived from stem cells can provide models of disease processes that then could be attacked under well-defined in vitro conditions. In a different type of presentation, Alon Chen provides a vision of how stem cell biology could be applied in a neuroendocrine system crucial for responses to stress: the corticotropin-releasing hormone system.

The final chapter, from the highly experienced developmental biology lab of Karine Rizzoti and Robin Lovell-Badge at the Crick Institute, presents an overview from both outside and inside the central nervous system of the likely contributions of such work to the new field of regenerative medicine.

New York, NY, USA
Boulogne Billancourt, France

Donald Pfaff
Yves Christen

Acknowledgments

The editors wish to express their gratitude to Mrs. Mary Lynn Gage for her editorial assistance and Mrs. Astrid de Gérard for the organization of the meeting.

Contents

List of Contributors

Cynthia L. Andoniadou Division of Craniofacial Development and Stem Cell Biology, King's College London, London, United Kingdom

Seth Blackshaw Solomon H. Snyder Department of Neuroscience, Johns Hopkins University School of Medicine, Baltimore, MD, USA

Institute for Cell Engineering, Johns Hopkins University School of Medicine, Baltimore, MD, USA

Department of Ophthalmology, Johns Hopkins University School of Medicine, Baltimore, MD, USA

Department of Neurology, Johns Hopkins University School of Medicine, Baltimore, MD, USA

Center for Human Systems Biology, Johns Hopkins University School of Medicine, Baltimore, MD, USA

A. Borsini Department of Psychological Medicine, Section of Stress, Psychiatry and Immunology, King's College London, Institute of Psychiatry, Psychology and Neuroscience, London, United Kingdom

Alon Chen Department of Stress Neurobiology and Neurogenetics, Max Planck Institute of Psychiatry, Munich, Germany

Department of Neurobiology, Weizmann Institute of Science, Rehovot, Israel

Jacques Drouin Laboratoire de Génétique Moléculaire, Institut de Recherches Cliniques de Montréal (IRCM), Montréal, QC, Canada

Daniel A. Lee Solomon H. Snyder Department of Neuroscience, Johns Hopkins University School of Medicine, Baltimore, MD, USA

Division of Biology and Biomedical Engineering, California Institute of Technology, Pasadena, CA, USA

Robin Lovell-Badge The Crick Institute, Mill Hill Laboratory, The Ridgeway, London, United Kingdom

Maryem Manzoor The Rockefeller University, New York, NY, USA

Thomas Pak Solomon H. Snyder Department of Neuroscience, Johns Hopkins University School of Medicine, Baltimore, MD, USA

D. Pfaff The Rockefeller University, New York, NY, USA

Carlotta Pires University of Copenhagen, Frederiksberg, Denmark

Karine Rizzoti The Crick Institute, Mill Hill Laboratory, The Ridgeway, London, United Kingdom

Joel N.H. Stern Department of Neurobiology and Behavior, The Rockefeller University, New York, NY, USA

Department of Autoimmunity, The Feinstein Institute for Medical Research, Northwell Health System, Manhasset, NY, USA

Departments of Neurology, Molecular Medicine, and Science Education, Hofstra Northwell School of Medicine, Hempstead, NY, USA

Lorenz Studer Developmental Biology, The Center for Stem Cell Biology, Memorial Sloan-Kettering Cancer Center, New York, NY, USA

Hidetaka Suga Department of Endocrinology and Diabetes, Nagoya University Hospital, Nagoya, Aich, Japan

Inna Tabansky Department of Neurobiology and Behavior, The Rockefeller University, New York, NY, USA

Viviane Tabar Department of Neurosurgery, The Center for Stem Cell Biology, Memorial Sloan-Kettering Cancer Center, New York, NY, USA

Hugo Vankelecom Department of Development and Regeneration, Cluster of Stem Cell Biology and Embryology, Unit of Stem Cell Research, KU Leuven (University of Leuven), Campus Gasthuisberg O&N4, Leuven, Belgium

Sooyeon Yoo Solomon H. Snyder Department of Neuroscience, Johns Hopkins University School of Medicine, Baltimore, MD, USA

P.A. Zunszain Department of Psychological Medicine, Section of Stress, Psychiatry and Immunology, King's College London, Institute of Psychiatry, Psychology and Neuroscience, London, United Kingdom

A Brief Overview of Techniques for Modulating Neuroendocrine and Other Neural Systems

Maryem Manzoor and Donald Pfaff

Abstract The history of experimental approaches to the nervous system forms the backdrop for new opportunities of using stem cell technologies in neuroendocrine systems. The emphasis of this chapter is on attempts at therapeutic maneuvers.

A Brief View of the Oldest, Most Primitive Approaches

No one uncovers the historical roots, the origins of ancient neuroscience, better than Stanley Finger of the Washington University School of Medicine. Egyptians whose names have been lost, writing during the age of the Pyramids, treated "involved individuals who suffered from head injuries. The descriptions revealed that early Egyptian physicians were aware that symptoms of central nervous system injuries could occur far from the locus of the damage." The Greek physician Alcmaeon (around the fifth century BCE) did various dissections and "proposed that the brain was the central organ of sensation and thought." But things got serious when the Greek anatomist Galen (AD 130–200) numbered the cranial nerves, distinguished sensory and motor pathways, distinguished the cerebellum from the cortex, and described the autonomic ganglia that control our viscera.

The historical origins of the information on sensory pathways begin with studies of the visual system that "described two distinct types of endings (rods and cones) in the retina" and later, in fact, the discovery of one of the visual pigments, rhodopsin. Anatomical studies then proceeded to the visual pathways, both the direct "reflex action" pathways to the superior colliculus and to the classical thalamo-cortical system. In turn, one contrasts vision with olfaction, which does not use the thalamus to signal to the cortex. According to Finger, "until the second half of the eighteenth century, air was viewed as an element and passive carrier of foreign particles that could affect the health of an organism." Putrid smells were avoided. Soon the adequate stimulus for olfaction as particles in the air was

M. Manzoor • D. Pfaff (✉)

The Rockefeller University, New York, NY, USA

e-mail: pfaff@mail.rockefeller.edu

© The Author(s) 2016

D. Pfaff, Y. Christen (eds.), *Stem Cells in Neuroendocrinology*, Research and Perspectives in Endocrine Interactions, DOI 10.1007/978-3-319-41603-8_1

1

recognized. It was known that olfactory receptors were in the nose, but the exact locations of the receptor-bearing cells were not known until the end of the nineteenth century. As with vision, investigations then proceeded to the central olfactory pathways in the brain.

Some of the initial findings reported paralysis on the side of the body that was opposite to brain damage that was limited to the cerebral cortex. Theorists supposed that the motor cortex was toward the front of the brain. But, in Finger's words, the "unequivocal experimental confirmation of a 'motor' cortex' electrically stimulated that part of cortex and caused movement." Confirming their results, subsequent removal of that part of the cortex of laboratory dogs led to motor deficits. Then, neurophysiologists would go on to define motor cortex precisely and to describe the motor tracts that lead from the forebrain toward the spinal cord.

Early ideas about emotion emphasized our visceral nervous systems, including both the sympathetic nervous system (raising blood pressure, heart rate, etc.) and the parasympathetic nervous system (usually the opposite effects of the sympathetic). In fact, the great psychologist/philosopher William James, at the end of the nineteenth century, actually proposed that we feel emotions consequent to changes in the autonomic nervous systems—feelings secondary to vascular changes. Walter Bradford Cannon and Philip Bard (at Johns Hopkins University) took a more straightforward view because they were able to stimulate the hypothalamus and directly cause emotional changes in experimental animals, changes like the induction of rage behavior. In subsequent years, the circuitry of the forebrain connected intimately to the hypothalamus (where we have done most of our work) proved to be essential for the performance of all emotional and motivated behaviors.

In the nineteenth century, clinicians had to deduce "how the brain works" by observing how behavioral capacities changed after brain damage. A special case was the British neurologist John Hughlings Jackson, who inferred which brain centers were "higher" and which "lower" by carefully noting how certain epileptic seizures in a given patient changed across time.

The Disciplines

Historically, the temporal order of accomplishment and understanding of brain function was that structure (morphology) led the way, followed by physiology (electrical recording), then chemistry (neurotransmitters), and now both genetics and computational neuroscience (in the most recent 30 or so years).

Morphology

Following Galen, mentioned above, an excellent example of progress comes from the work of Andreas Vesalius (1543, *De Humani Corporis Fabrica*). Here is the level of detail he achieved: "Professors of dissection usually divide the anterior brain, which they call the cerebrum, from the posterior brain, which they call the cerebellum: in turn, the anterior is normally divided into right and left. Not that the great masters of anatomy think that the brain is *entirely* divided..."

For me, the breakthrough to modern neurobiology occurred when chemical stains were discovered that would reveal microscopic details of nerve cells. The Italian scientist Camillo Golgi got a lucky break when nighttime cleaning personnel, servicing the hospital kitchen that Golgi had turned into a laboratory, knocked one of his human brain specimens into a slop bucket. Intrigued by the apparent staining of cells in that specimen, Golgi found that a key ingredient in turning some of the neurons dense-black was (and still is) silver nitrate. A brilliant exponent of Golgi stain-based nerve cell biology was the Spanish neuroanatomist Ramon y Cajal.

Cajal clearly stated the "neuron doctrine." The brain is not just a continuous string of fibers forming anastomoses to make never-ending nets. Instead, as Nobelist Cajal concluded, each nerve cell is an autonomous unit. "The neuron is the anatomical and physiological unit of the nervous system." And the rest is history. How do neurons talk to each other (Kruger and Otis 2007)? The Nobel winning physiologist Sir Charles Sherrington (1857–1952) "developed the concept of the synapse" and introduced modern neurophysiology in his 1932 book, "The Integrative Action of the Nervous System."

For decades the development and use of new neuroanatomical techniques dominated the scene. For example, a Dutch neuroanatomist, Walle J.H. Nauta, my teacher, who had survived World War II by eating tulip bulbs, came to the United States (MIT) and developed techniques for seeing very fine nerve fibers. This type of technical development led to our current state, when neuroscientists ambitiously are trying to map all the connections in the human brain.

Physiology

After microscopic techniques for looking at neurons gave our field a running start, scientists good at electrical recording invented what is called "neurophysiology." For example, in Britain, Lord Adrian received the Nobel Prize for showing, in 1938, how to record from individual nerve fibers. Later, tiny wire probes called microelectrodes were developed so that we could put them deep into the brain and record the electrical activity of individual neurons. Most prominent during the early years of this technical endeavor were David Hubel and Torsten Wiesel, who used such electrodes to elucidate the neurophysiology of the visual cortex. And, of course, recording in a non-invasive manner on the surface of the skin over the skull gives you "EEG:" electroencephalography of wave-like activity of the cerebral cortex so useful for clinical diagnosis, as in epilepsy or sleep problems. Sakmann's and Neher's Nobel prize-winning invention led to a modern development of the microelectrode: a tiny pipette that suctions onto the surface of an individual neuron, breaks through that membrane and records from inside the neuron. This is the "patch clamp" technique, which unveils the subtlest details of cross-membrane currents in nerve cells, especially in brain slices or in nerve cell culture.

Chemistry

Later still came the origins of neurochemistry. Of course, the discoveries of how neurotransmitters such as dopamine and acetylcholine are produced in neurons and how they are released at synapses and eventually broken down took center stage. The Nobel prize winner Julius Axelrod, running a large lab at the National Institutes of Health, became famous not only for his own work but also for mentoring an entire generation of neurochemical geniuses. One of those geniuses, Solomon Snyder, not only discovered opiate receptors in the brain but also could claim such a large number of advances in neurochemistry that the entire department of neuroscience at Johns Hopkins Medical School now is named after him.

Rita Levi-Montalcini's discovery of nerve growth factor (NGF) opened a new arena of neurochemistry in which peptide chemistry was paramount and led to the elucidation of families of related growth factors.

As DNA's chemistry and its regulation in gene expression became easier and easier to study, neuroscientists jumped on the bandwagon. For example, I was able to prove (reviewed in Pfaff 2002; Lee et al, 2009) that expression of a particular gene (that which codes for an estrogen receptor) in particular neurons of the brain (hypothalamic and preoptic neurons) is absolutely essential for specific instinctive behaviors (mating behavior and maternal behavior). And now the focus has shifted to the nuclear proteins that coat DNA in the neuron and regulate gene expression.

Genetics, Genomics

To manipulate gene expression in neuroendocrine cells, siRNA (small interfering RNA) was used to knock out a single gene (estrogen receptor-alpha) in specific neurons, which abolished all aspects of female reproductive behavior: in temporal order, lateral preoptic neurons (courtship behavior); ventromedial hypothalamic neurons (sex behavior; reviewed in Pfaff 2002); and medial preoptic neurons (maternal behavior; Ribeiro et al. 2012). These studies comprised a behaviorally relevant extension of nuclear hormone receptor chemistry in neuroendocrinology.

Cognitive Neuroscience

As recently as 70 years ago, studies that dealt with complex behaviors—psychology, personality, and so forth—were dismissed by some as "soft." The scientific qualities of accuracy and precision were doubted for those fields. But the field of cognitive science has come a long way. As things began to improve, some 100 years ago, scientific approaches to the behaviors of animals were split into two parts. One approach, called ethology, most popular in Europe, usually treated the natural behaviors of animals in their natural environments. Ethology was rooted in biology. The other approach, experimental psychology, was more popular in America. Derived from physics, experimental psychological studies would feature well-controlled experiments in the laboratory to answer specific, precisely worded questions or to test formal hypotheses. Both of these approaches could be applied to human subjects. Finally, most famously, the Viennese neurologist Sigmund Freud originated the psychodynamic theory of the human mind and brain, psychoanalysis.

Cognitive neuroscientists often united these studies of behavior with the various neuroscientific methodologies and techniques mentioned above. Historically, brain lesions and their behavioral analyses came earliest. Well known currently, for example, is the patient HM. The Canadian neurosurgeon William Scoville removed most of his hippocampus on both sides of his brain to prevent continuing epileptic seizures. Then the Canadian psychologist Brenda Milner documented his permanent loss of memory for recent events. In other studies, human language was emphasized, as summarized by Chatterjee and Coslett (2014).

Looking back, the first great victory regarding language was the observation by the French neurologist Broca that loss of a delimited region on the lower side of the left frontal lobe impaired the production of speech. On the other hand, damage to a cortical area farther posterior, near the juncture of the temporal lobe and parietal lobe, again on the left side, would impair, in Heidi Roth's words "the acoustic images of words." Patients with this type of brain damage, studied by the German neurologist Carl Wernicke, could not identify or recognize normal speech. As you can imagine, these studies were based on small numbers of patients. More patients had to be studied, brain damage had to be better defined and the language analyses had to be more sophisticated. But Broca and Wernicke had paved the way.

From there neurologists and neuroscientists went on to initiate the study of all aspects of human behavior. My own lab has zeroed in on the most fundamental influence within the brain, a concept I call "generalized brain arousal," which is essential for initiation of all behaviors. On the other hand, neurologists tend to concentrate on specific disorders, such as epilepsy, autism, memory, and addiction.

Computational Approaches

One branch of neuroscience came out of engineering, physics and mathematics. The theorem of McCulloch and Pitts, published in 1943, coupled with the interests of Alan Turing sparked the field alive. Then, in 1956, electrical engineers Claude Shannon (the inventor of information theory), John McCarthy and Marvin Minsky conceived and led the conference that generated the field of "artificial intelligence," the basis of sophisticated robotic behavioral regulation. All these scientists were applying techniques that had already been proven successful and using them for the potential understanding and mimicry of the brain's behavioral regulation.

Neuroscientific work now has reached such a level of precision that our data often can be treated with computations based on applied mathematics and statistics. Computational neuroscience can be divided into two parts: analysis and so-called "modeling," which means devising computer programs that are supposed to embody the essential features of some well-chosen groups of neurons in the brain. Both parts of computational neuroscience contribute to the type of artificial intelligence that regulates behaviors by robots and computations by neural networks.

As stated by Eve Marder, a prominent computational neuroscientist at Brandeis University, "computational models are invaluable and necessary in this task and yield insights that cannot otherwise be obtained. However, building and interpreting good computational models is a substantial challenge, especially so in the era of large datasets." Fitting detailed models to experimental data is difficult and often requires onerous assumptions, whereas more loosely constrained conceptual models that explore broad hypotheses and principles can yield more useful insights.

George Reeke, at Rockefeller University, envisions modeling of the brain as an obvious approach to answering questions all of us have about the brain: how are sensations, categorized, how are actions selected from a given repertoire, how is "motivation" to be conceived (2012)? It was the availability of computers that allowed academic researchers to construct ever more detailed and complicated models of the brain. Some neural modelers try actually to mimic neurons and neuronal systems faithfully, in detail, while others do not; instead, in Reeke's words, they just concentrate on devising "rule-based systems." In all cases, the equations neuronal modelers use to mimic neurons never match the full sophistication and flexibility of real neurons; neither are the circuitry properties of the human brain truly realized, even in the best models. These days, many neuroscientists are drawn into modeling, and thus a form of AI, because of the considerable number of free software modeling packages. The implication is that progress in neuronal modeling is accelerating. Nevertheless, as Reeke points out, the field is not without its shortcomings. For example, in some cases, the equations representing neurons and their connections are so abstract that they lose the properties of real neural systems. In other cases, neuronal modelers will run large numbers of trials and select some in which their favorite ideas work, which, of course, leads to false conclusions.

The operations of individual nerve cells and individual synapses comprise the irreducible base of neuronal modeling and have absorbed the attention of William Lytton, at State University of New York Medical Center. One starts with the nerve cell membrane. The equations that represent the membrane in the model contain the elements of electrical circuit theory: resistors and capacitors. Once those equations and the dynamic changes when electrical current flows, for example through sodium channels or calcium channels, are in place, you are ready to start building artificial "circuits." One example would be the modeling of a type of connection serving the passage of sensory information through the thalamus with its subsequent impact on the cerebral cortex, which can be modeled, as Lytton has done, using five types of "neurons" and nine types of connections between neurons.

A Brief Survey of Emerging Techniques for Neuromodulation

Electrical

One striking development demonstrated the use of the patient's own electrical waveform activity to move artificial limbs. While John Donoghue (Brown) was given a lot of credit for opening up this field, Miguel Nicolelis (Duke, Sao Paolo) has reported similar achievements. Dedicated to the use of helping injured war veterans with artificial limbs, Geoffrey Ling (DARPA) has shown effective control of artificial limbs in therapeutic settings.

On the sensory side of the CNS, some scientist/engineers are concerned with age-related macular degeneration. Retinal prostheses to help ameliorate this problem are a central concern of Sheila Nirenberg (Cornell), but the project has a dimension that goes well beyond prosthesis construction. Central to the solution is a deep understanding of the critical features of electrical signaling to the optic nerve. Working with the computational neuroscientist Jonathan Victor, Nierenberg is discovering the answer to that intellectual problem now.

Chemical

While the field of microfluidics has been applied extensively to sampling extremely small volumes of biological fluids, it will now become available for precise, time-limited local delivery of therapeutic substances in specific brain regions. For similar purposes, nanoparticles, lipid bilayered to cross blood–brain barrier, can be loaded up with chemicals intended for therapeutic purposes. Cationic liposomes, the positive charge offering the possibility of entry into cells, can be used likewise.

DREADDS—Designer Receptors Exclusively Activated by Designer Drugs—can be genetically encoded so that they are expressed only in specific subpopulations of neurons, thus to bind pharmaceuticals..

Genetic

The applicability of optogenetics to the nervous system (Karl Deisseroth, Stanford) has been proven; it uses brief pulses of light to activate channel proteins that, in some cases, excite neurons and, in other cases, inhibit neurons. For example, inhibiting GABA neurons that, in turn, inhibit giant medullary reticular neurons can enhance recovery of consciousness from anesthesia, as measured by behavioral activation and by the activation of the cortical EEG.

Viral

Locally delivered by stereotaxically guided microinjection, adeno-associated viral particles (AAV) are outfitted with cell-selecting promoters to modify synthetic and electrical activities of selected subsets of neurons in that neuronal group (only).

Computational

In general, the use of temporal and spatial patterns of firing in the human brain's "connectome" requires big data computational efficacy. One specific example, viewing brain activity as a set of non-linear dynamic systems, would involve the identification and use of "attractor" states of neuronal circuitry. This project is being carried out in the context of the Obama BRAIN initiative.

Special Opportunities for Manipulating the Unique Products of Neuroendocrine Neurons

Because neuroendocrine neurons specifically produce small chemicals of surpassing importance for the governance of the physiology of the entire body, the possibility of using chemical, viral or genetic means to regulate their activity offers unique therapeutic opportunities. Seven examples:

- GnRH: Gonadotropin releasing hormone controls all of reproductive physiology and reproductive behavior.
- GHRH: Growth hormone releasing hormone promotes the release of growth hormone from the pituitary.
- Somatostatin: Reduces the release of growth hormone from the pituitary.
- TRH: Thyrotropic releasing hormone facilitates the release of TSH from the pituitary Normal mentation and mood depend on thyroid hormone levels.
- CRH: Corticotropic releasing hormone facilitates the release of ACTH from the pituitary (stress response). In the brain, CRH (also known as CRF) participates in circuits that govern stress-related behaviors.
- Oxytocin: In addition to regulating lactation and parturition, oxytocin participates in the initiation of maternal behavior and prosocial motivation.
- Vasopressin: Regulation of body water, blood pressure, blood volume. Vasopressin expression in certain forebrain neurons is known to facilitate aggression.

This Volume

As illustrated throughout this volume, stem cell biology is a fast-moving, young field with obvious therapeutic potential as well as technical and legal encumbrances. Following a didactic chapter intended for readers without a background in this area of medical science, several chapters will report striking advances. Between invitations for this project and the time of the meeting in Paris, one doctor moved from having a definite need for stem cell technology to publishing a pair of papers in the prestigious journals, *Science* and *Cell*. The meeting wrapped up with an overview from the Lovell-Badge group, the lab that 24 years ago reported the basis of sexual differentiation through the discovery of the SRY gene on the Y chromosome.

Further Reading

Catani M, Sandrone S (2015) Brain renaissance: from Vesalius to modern neuroscience. Oxford University Press, Oxford

Chatterjee A, Coslett HB (2014) The roots of cognitive neuroscience. Oxford University Press, Oxford

Dertouzos M et al (1974) Systems, networks and computation. McGraw-Hill, New York

Finger S (1994) Origins of neuroscience. Oxford University Press, Oxford

Gagnidze K, Weil ZM, Faustino LC, Schaafsma SM, Pfaff DW (2013) Early histone modifications in the ventromedial hypothalamus and preoptic area following oestradiol administration. J Neuroendocrinol 10:939–955

Hodges A, Turing A (1983) The enigma. Princeton University Press, Princeton

Kruger L, Otis TS (2007) Whither withered Golgi? A retrospective evaluation of reticularist and synaptic constructs. Brain Res Bull 72:201–207

Lee A et al (2009) In: Pfaff D (ed) Hormones, brain and behavior, 2nd edn. Academic Press, Elsevier, San Diego

Lytton W (2012) In: Pfaff D (ed) Chapters in neuroscience in the 21st century. Springer, Heidelberg

O'Leary T, Sutton AC, Marder E (2015) Computational models in the age of large datasets. Curr Opin Neurobiol 32C:87–94

Pfaff D (ed) (2002) Hormones, brain and behavior, 1st edn. Academic Press, Elsevier, San Diego

Pfaff D, Joels M (eds) (2016) Hormones, brain and behavior, 3rd edn. Elsevier, Cambridge

Pfaff D (2012, 2015) (ed) Neuroscience in the 21st century (a five volume text free in poor countries, 1st edn. Heidelberg, Springer (Pfaff D, Volkow N (eds) 2nd edn. Heidelberg, Springer)

Reeke G (2012) In: Pfaff D (ed) Neuroscience in the 21st century. Springer, Heidelberg

Ribeiro AC, Musatov S, Shteyler A, Simanduyev S, Arrieta-Cruz I, Ogawa S, Pfaff DW (2012) siRNA silencing of estrogen receptor-α expression specifically in medial preoptic area neurons abolishes maternal care in female mice. Proc Nat Acad Sci USA 109(40):16324–16329

Wilson EG (2006) The melancholy android. SUNY Press, Albany

Basics of Stem Cell Biology as Applied to the Brain

Inna Tabansky and Joel N.H. Stern

Abstract Stem cell technology can allow us to produce human neuronal cell types outside the body, but what exactly are stem cells, and what challenges are associated with their use? Stem cells are a kind of cell that has the capacity to self-renew to produce additional stem cells by mitosis, and also to differentiate into other—more mature—cell types. Stem cells are usually categorized as multipotent (able to give rise to multiple cells within a lineage), pluripotent (able to give rise to all cell types in an adult) and totipotent (able to give rise to all embryonic and adult lineages). Multipotent adult stem cells are found throughout the body, and they include neural stem cells. The challenge in utilizing adult stem cells for disease research is obtaining cells that are genetically matched to people with disease phenotypes, and being able to differentiate them into the appropriate cell types of interest. As adult neural stem cells reside in the brain, their isolation would require considerably invasive and dangerous procedures. In contrast, pluripotent stem cells are easy to obtain, due to the paradigm-shifting work on direct reprogramming of human skin fibroblasts into induced pluripotent stem cells. This work has enabled us to produce neurons that are genetically matched to individual patients. While we are able to isolate pluripotent stem cells from patients in a minimally invasive manner, we do not yet fully understand how to direct these cells to many of the medically important neuroendocrine fates. Progress in this direction continues to be made, on multiple fronts, and it involves using small molecules and proteins to mimic developmentally important signals, as well as building on advances in "reprogramming" to directly convert one cell type into another by forced expression of sets of transcription factors. An additional challenge involves providing these cells with the appropriate environment to induce their normal behavior

I. Tabansky (✉)
Department of Neurobiology and Behavior, The Rockefeller University, New York, NY, USA
e-mail: inna.tabansky@gmail.com

J.N.H. Stern
Department of Neurobiology and Behavior, The Rockefeller University, New York, NY, USA

Department of Autoimmunity, The Feinstein Institute for Medical Research, Northwell Health System, Manhasset, NY, USA

Departments of Neurology, Molecular Medicine, and Science Education, Hofstra Northwell School of Medicine, Hempstead, NY, USA

© The Author(s) 2016
D. Pfaff, Y. Christen (eds.), *Stem Cells in Neuroendocrinology*, Research and Perspectives in Endocrine Interactions, DOI 10.1007/978-3-319-41603-8_2

11

outside the body. Despite these challenges, the promise of producing human neuroendocrine cell types in vitro gives opportunities for unique insights and is therefore worthwhile.

Introduction

By the beginning of the twentieth century, humanity knew that the basic unit of the brain was the neuron. We also knew that a person was born with all the neurons she would ever have, as these neurons could not—under any conditions—regenerate. This understanding left patients with diseases resulting from neuronal death caused by injury or autoimmunity with few options. Over the course of the twentieth century, this dogma has been overturned, driven by two advances: (1) the discovery of neural stem cells, and (2) reprogramming technology that allows us to make neurons that are genetically matched to individual people outside the body. While the opportunities are clear, considerable technical challenges remain before they can be fulfilled in the clinic.

The Basic Biology of Stem Cells

A stem cell is defined as any cell type with two fundamental capacities (1) self-renewal and (2) differentiation. Self-renewal refers to a cell's capacity to divide and make other cells with the same properties. Differentiation refers to its ability to make other cell types, performing other biological functions.

For instance, hematopoietic stem cells are found in the bone marrow, where they generate progenitor cells that give rise to the cells of the immune system and red blood cells.

Not all stem cells have the same "potency," the capacity to give rise to similar cell types. Broadly speaking, they are characterized as totipotent, pluripotent and multipotent. The hematopoetic stem cells mentioned earlier are a multipotent cell type: they are able to give rise to many kinds of cells, but only of the blood lineage.

In basic embryology, blood originates from the mesoderm, the middle layer of an embryo, which forms as the embryo undergoes a process called "gastrulation" shortly after fertilization. Gastrulation subdivides the cells in the group into three broad layers: endoderm, which gives rise to the cells of many internal organs, mesoderm, which gives rise to the muscles and the blood, and ectoderm, which gives rise to the nervous system and epithelial layers. These three lineages are referred to as the embryonic germ layers.

For mammals, even before gastrulation occurs, the tissues of the embryo are classified into two other broad categories: extra-embryonic and embryonic.

Extra-embryonic tissues are "outside the embryo," referring primarily to the cells of the amniotic sac and the placenta: organs that are essential for embryonic development but are discarded after birth.

To be classified as "multipotent," stem cells must make at least two different lineages, usually from the same embryonic germ layer. In contrast, pluripotent stem cells can make multiple lineages from all three embryonic germ layers but not from extra-embryonic tissue. Finally, totipotent stem cells can make all three embryonic germ layers and the extra-embryonic tissue. The only known indisputably totipotent cell is the zygote.

Preimplantation Development and Embryonic Stem (ES) Cells

In most animals, development occurs outside the body and the embryo is not physically connected to the mother. Mammals, particularly placental mammals, are an exception. However, even in placental mammals, not all development occurs in the uterus. During the first few days of its development (exact number of days varies depending on the species), the early mammalian embryo travels down the fallopian tubes into the uterus. Once inside the uterus, the embryo invades the uterine wall and establishes the organs that will support its further development—a phenomenon known as implantation. Thus, the first days of development within the fallopian tubes are called "preimplantation development."

During preimplantation development, several important developmental events occur. Of the biggest relevance to us is the first cell fate determination, or segregation of the early totipotent cells into two lineages: extra-embryonic and embryonic.

We will review these events as they occur in the mouse, the most commonly studied mammalian model of development, and then discuss differences between human and mouse development. At the first stage of development, the fertilized zygote undergoes a series of three cell divisions to produce eight cells. At these early stages, these cells are called blastomeres. The divisions that produce these blastomeres are thought to be mostly "symmetric" (to produce cells with similar properties), though blastomeres have been reported to exhibit bias toward particular developmental lineages (Tabansky et al. 2013). During these early divisions, cells do not increase in size: every division produces two daughter cells that are half the size of the mother; they are called "cleavage" divisions.

Until the eight-cell stage, these cleavage stage blastomeres have very few cell adhesion molecules, and they are separate from each other and readily distinguishable under a microscope. However, at the eight-cell stage, the molecules on the cell membrane start to bind to each other, and the boundaries of the cells become indistinguishable. This moment in development is called "compaction," and though compaction is morphologically striking, it is far from being a mere cosmetic change. Instead, it serves a very important role: differentiating the inside of the embryo from the outside for the first time.

Immediately after compaction, most of the blastomeres are still able to give rise to embryonic and extra-embryonic lineages. However, as they continue to divide, some cells become separated from the outside. At the same time, the tight junctions between the outside cells allow the formation of a fluid-filled cavity within the embryo. The cells on the outside will now comprise the trophectoderm, which gives rise to the placenta. Inside of the fluid-filled cavity, known as the blastcoel, the cells with no contact with the outside of the embryo form a clump that adheres to the trophectodermal cells. This clump is known as the inner cell mass (ICM). It contains the pluripotent cells that will give rise to the embryo proper, as well as a newly formed cell lineage that will give rise to the amniotic sack: the primitive endoderm, or PE.

The trophectoderm is the cell lineage that will intercalate with the uterine lining and allow implantation to occur. As this process proceeds, the pluripotent lineage loses its ability to form PE, becoming another cell type known as the epiblast. The distinction between ICM and epiblast is very important for understanding the differences between mouse and human embryonic stem cells.

Derivation and Maintenance of Pluripotent Stem Cells: Differences Between Mouse and Human

Mouse ES cells have been known and used for years before human embryonic stem cells were derived (Thomson 1998).

While mouse and human ES cells indubitably share multiple features, including pluripotency and the capacity to self-renew, they do not grow under the same conditions in culture. More specifically, mouse ES cells absolutely require activation of the JAK-STAT3 signaling pathway in order to continue to proliferate, usually achieved by the addition of the Leukemia Inhibitory Factor (LIF) to the medium. In contrast, human ES cells absolutely require basic fibroblast growth factor (bFGF) and Activin A signaling, and they will lose their ability to differentiate and grow without them.

Human ES cells are not unique: ES cells isolated from most species share the features of human ES cells, but not mouse ES cells. The question then becomes, why is the mouse the outlier?

Mice have a unique property known as diapause; in times of stress or starvation, females can delay implantation of blastocysts, which persist in the oviduct until conditions improve. Most mammals do not have this ability. Diapause is mediated by LIF; in fact, defects in diapause are the main phenotype of LIF-knockout mice. These observations led researchers to conclude that conditions for culturing mouse ES cells mimic the response of the ICM to diapause, whereas the conditions for culture of ES cells from other animals do not.

However, this raises an important point: why is it possible to derive ES cells from other species at all? The currently favored hypothesis suggests that most ES cultures mimic the conditions that exist in the embryo a little after implantation but before the cells have begun the process of migration that will separate them into the three germ layers. At this stage, the pluripotent lineage is called the epiblast, and the cells derived from it can therefore properly be called "epiblast stem cells."

The hypothesis described above makes several predictions about the nature of human and mouse ES cells. One is that they will require different conditions and display different properties. Indeed, they do: mouse ES cells have different growth requirements, different differentiation requirements and different morphology than human ES cells.

A second prediction would be that, if differences between mouse and human ES in fact reflect different developmental states, then it should be possible to derive mouse ES cells that have a more human-like phenotype, growth factor requirement and morphology. Indeed, mouse epiblast stem cells were derived a few years ago, and they share many of the characteristics of the cell type known as human ES cells (Tesar et al. 2007). Mouse ES cells can also be converted to mouse epiblast cells, and vice versa (Greber et al. 2010).

These findings have multiple applications for stem cell research. Of these, perhaps the most urgent is that testing protocols on cheaper mouse ES cells before trying them on human ES cells is not a good idea, as mouse ES cells are fundamentally different and respond to differentiation cues in a manner highly dissimilar to human ES cells. However, it is possible to test differentiation protocols on mouse epiblast stem cells as they respond to differentiation cues in a manner quite similar to human ES cells.

How to Test Pluripotency?

The definition of stem cells is primarily functional. Therefore, any test to determine whether a stem cell is in fact a stem cell must also be functional. For pluripotent stem cells, this functionality encompasses the ability to self-renew and also to differentiate into any cell type in the body.

The first property is quite easy to test: simply assess whether stem cells continue to grow and produce more pluripotent stem cells. However, how do you test whether a cell can differentiate into anything in the body?

In mouse ES cells, there are two tests of increasing stringency. In the less stringent version of this test, pluripotent ES cells are injected back into the cavity of the blastocyst, where they aggregate with the inner cell mass and, ideally, contribute to the germ line and multiple other lineages. Usually the coat color of the "recipient" blastocyst into which the cells were injected is different than the color of the original "donor" mouse from which the stem cells were derived. The chimeric mice therefore have variegated coloring resulting from a mix of two cells of two different genotypes in their skin.

This technology is also used to make transgenic mice: stem cells are genetically modified in an appropriate way, and the chimeric mice resulting from the stem cell transfer into the blastocyst are crossed to a wildtype mouse. If the stem cells contributed to the germ line of the chimera, these animals can be expected to produce at least some progeny where every cell carries the transgene. The presence of the transgene in these progeny animals can be assessed by analyzing DNA from their skin cells.

A more stringent test of pluripotency in mice relies on the fact that the embryo has a form of quality control where only cells with two copies of the genome (one from the mother, one from the father) can contribute to the adult organism. In using this approach, people wait for the first division of the recipient embryo and then fuse the two cells back together into one cell. The embryo continues to develop to the blastocyst stage, but each of its cells now contains four copies of its genome: two from the father and two from the mother, a feature called being "tetraploid." Due to that feature, the cells in the embryo are only able to form the placenta and other extra-embryonic lineages and cannot contribute to the adult. However, if the pluripotent cells with the normal number of genomes are introduced into this embryo, they will form all the lineages of the adult. Because they are complementing the function that the tetraploid cells lost in embryonic development, this technique is known as "tetraploid embryo complementation." It is considered the gold standard of pluripotency in the mouse, but it can also be used to generate transgenic mice more quickly.

However, neither of these techniques is applicable to humans, due to both technological and ethical reasons. Therefore, the test for pluripotency in human cells must be something different and less stringent.

One simple test is to remove the bFGF—on which the human ES cells rely to stay pluripotent—from the media and to allow the cells to differentiate without trying to influence their path. This test is frequently used as a preliminary characterization of newly derived human pluripotent cell lines.

A more stringent test is to implant the cells into the body cavity of an immunocompromised mouse, where they will continue to grow, giving rise to a tumor, called a teratoma, containing multiple fully differentiated lineages. After the tumor grows, it is possible to test the number of different cells that were able to develop within the mouse.

Why not just carry the whole test out in a dish, instead of implanting into a mouse? Different cell types need different environments to grow, and it is impossible to combine them all in the same preparation of cells and to allow them to survive until analysis. However, in the mouse, the supply of blood and oxygen from the body allows the teratoma to develop in a manner somewhat similar to what might happen in an embryo, but in a more disordered fashion. Since the environment is more supportive of multiple different cell types, more different kinds of cells in more mature states can be detected and the test is more stringent.

It is worthwhile mentioning that there are vast numbers of different kinds of cells within the body, and it would be a daunting task to attempt to detect them all within a teratoma. Therefore, while the teratoma can detect the ability of a cell to give rise to all three germ layers, it cannot be used as evidence that a particular cell line can give rise to every single kind of cell in the body. Thus, a teratoma is an approximation of a test for the most stringent definition of pluripotency.

Opportunities and Challenges for Using ES Cells in Medicine

What do we do with pluripotent stem cells once we have them? Multiple uses have been proposed for these cells, including (1) studying rare cell types, (2) disease modeling, (3) drug screening and (4) transplantation therapies.

Of these, the most obvious and simple application is studying rare cell types. While mice are readily accessible and their neurons can easily be isolated from the brain and cultured in a dish, human cells are not always so easy to isolate and manipulate. This is especially true in the brain, as death is currently primarily defined by the cessation of brain function. Therefore, unlike many cells, neurons cannot be harvested from people who have opted to donate their organs to research, as the damage to the brain that is necessary to declare a person dead will also affect the cells.

To study human neurons in detail another source of cells must be found, and neurons derived from human pluripotent cells constitute one such source. Pluripotent cells from most species tend to be predisposed to make neurons, making such neurons easy to obtain. Additional protocols have been developed to ensure that particular kinds of cells—of interest to people from the investigation of diseases perspective—are preferentially made (Tabar and Studer 2014).

Growing neurons in culture can and has been used to address many questions about their basic biology and their electrophysiological properties. However, it is also true that results from experiments on cultured neurons need to be interpreted very carefully. This caution should particularly apply to human neurons when they are being studied outside the body and when differences between human and mouse are revealed by the study. The question will always arise whether the differences observed have to do with something that happens in the human brain or whether they arise from the distinct ways that human and mouse neurons adapt to the environment outside the body. Luckily, if the biochemical basis of the phenomenon is known, the neurons in culture can be compared to human postmortem brains to determine whether the phenomenon under study occurs in the body as well as in cell culture.

Disease modeling builds on the study of normal human cell types, by comparing cells that are obtained from pluripotent stem cells of patients with a particular (usually genetic) disorder with cells from patients who do not have this particular disorder. Prominent examples include amyotrophic lateral sclerosis and schizophrenia (Marchetto and Gage 2012). Disorders where a person with a particular genetic makeup is highly likely to get a disease are easier to study in culture than disorders that develop in response to environmental stimuli or involve multiple cell types, such as autoimmune diseases or Alzheimer's disease. However, when the cells are provided with the proper environmental stimuli to induce a disease-like state, it may eventually be possible to model a wide range of diseases in culture.

Once a good disease model has been established, drug screening can begin. Drug screening in culture builds on disease modeling by treating cells with various potentially therapeutic compounds and attempting to determine which compounds can reverse or slow down the course of the disease. The simplest approach is to use cells that express some sort of fluorescent protein or that secrete a particular metabolite that indicates health and then measure how treatment with compounds can alter the amount of fluorescence (a proxy for cell number) or metabolite in the dish. Automated drug screening robots that can measure fluorescence from tens of thousands of different samples are routinely used for drug screening. In this case, it is not even necessary to know the mechanism of disease or the mechanism of action of the compound in order to isolate an effective drug; however, it is desirable to understand at least a little about the function of the drug before administering it to patients.

Of all these approaches to using stem cells for medicine, perhaps the most daunting and fraught with potential side effects is transplantation of stem cells and cells derived from them back into a patient. Ideally, the cells would be perfectly genetically matched to the patient, negating the necessity for immunosuppressive drugs, which are necessary for conventional organ transplantation. This approach can be risky because cells tend to accumulate abnormalities in culture, potentially causing some of them to become tumorigenic; also, if the pluripotent cells are insufficiently differentiated, their inherent tumorigenicity (see above) also becomes a problem. However, recent phase 1 clinical trials have at least suggested that stem cells could potentially cause functional improvements—with few adverse effects— over the course of several years (Schwartz et al. 2012); whether this will hold true for larger cohorts and longer term trials remains to be determined.

Obtaining Cells Genetically Matched to Patients: Reprogramming, Cloning, and Induced Pluripotent Stem Cells

In animals, pluripotent stem cells can be derived from embryos quite easily, but human preimplantation embryos, while sometimes used in research in very specific circumstances, are not widely available. In addition, the cells used for modeling disease, drug screening and transplantation need to be genetically identical to the patient, necessitating that the cells be derived from the person and not from their offspring.

Given that a patient is an adult and therefore does not have any more embryonic cells, some applications require that the cells be induced to revert back to an embryonic-like state, or "reprogrammed." While here reprogramming refers to the conversion to an embryonic-like state, the term can also indicate a direct interconversion of two different cell types into each other: for instance, a muscle cell into a neuron. It generally refers to the types of interconversion that do not occur under natural circumstances. In contrast, the term "differentiate" refers to making a more adult cell type from a more embryonic cell type (or from a multipotent stem cell), thereby replicating a process that normally occurs in nature.

Historically, there have been three methods for obtaining pluripotent stem cells from patients: cell fusion, somatic cell nuclear transfer and direct reprogramming. Of these, cell fusion is the simplest technique. The cytoplasms of the cells are induced to combine together to form one cell (the nuclei can also combine into a single tetraploid nucleus). Interestingly, if cells of different type are fused, they do not produce an intermediate kind of cell. Instead, one of the cell types is "dominant" over the other, and the resulting cell will have multiple nuclei but will otherwise be functionally very similar, if not identical, to the dominant cell type. It so happens that pluripotent cells are dominant over every other kind of cell, allowing reprogramming by cell fusion.

However, the complication of this method is that, while it may theoretically be possible to enucleate one of the cells or to remove one of the nuclei after fusion, no practical method for doing so on a large scale has yet found wide acceptance. Thus, most products of cell fusion are tetraploid (with the associated problems) and, to make pluripotent stem cells genetically matched to a patient, you would have to start with pluripotent cells from that patient, which obviates the usefulness of the whole endeavor.

An alternate method of reprogramming cells to a pluripotent state first came into prominence in 1996, when people were able to produce an adult sheep from a skin cell isolated from another sheep. In this approach, called "somatic cell nuclear transfer" (SCNT) and referred to colloquially as "cloning," an egg cell has its nucleus removed and replaced with a nucleus from a donor cell. In a way, SCNT is simply a special case of cell fusion of an enucleated totipotent zygote with a differentiated cell. The egg cell then goes on to develop as though it is an embryo, producing a blastocyst from which stem cells can be derived and also, potentially, an adult animal. Blastocysts have been produced by SCNT from multiple animals, including, quite recently, humans (Chung et al. 2014). However, logistical and ethical considerations involved with obtaining human eggs and making embryos preclude this research from being applicable on a large scale to medicine. It may eventually be possible to make cells resembling human eggs in culture from pluripotent cells, but there are currently no established protocols for this approach.

Currently the most popular method of reprogramming for drug screening and disease modeling relies on the delivery of four transcription factors (genes that regulate expression of other genes) to adult cells in order to convert them into an embryonic-like state (Takahashi and Yamanaka 2006). Named after Shinya Yamanaka, who originally discovered this approach, they are also sometimes known as the "Yamanaka factors."

The original method relied on a type of genetically modified retrovirus from which the DNA encoding viral genomes was removed and replaced with DNA encoding each of the Yamanaka factors. Once the cell infected with the genetically modified virus became pluripotent, they were able to activate the intrinsic protective mechanisms found in pluripotent cells to inactivate this particular kind of virus. Thus, once reprogrammed, these cells were again differentiated into other cell types, and Yamanaka and multiple other groups were able to test the pluripotency of these cells. The caveat is that retroviruses by themselves are carcinogenic, and their presence is undesirable for any cells being transplanted back into patients, which is why the cells reprogrammed by the Yamanaka method (called induced pluripotent stem cells) are used primarily for disease modeling and studies of diseases processes. However, multiple groups have published papers on alternative approaches to reprogramming, including pieces of DNA that do not integrate into the genome, a special kind of RNA molecule called micro-RNA and small molecules (Schlaeger et al. 2015). The Yamanaka method currently remains the most widespread technique but, going forward, it is quite likely that one of these other methods will eventually replace it.

Opportunities and Challenges of Producing Hypothalamic Neurons from Stem Cells

There is wide agreement that investigation of the function of the human hypothalamus could be enhanced by the production of hypothalamic-like neurons from ES cells. Many diseases exist in which particular subpopulations of hypothalamic neurons are absent or defective, and replicating the disease in culture for testing of drug candidates or even producing the neurons and transplanting them back into patients are obvious therapeutic opportunities. However, before neurons can be investigated or transplanted, they first need to be produced, and that is quite a daunting challenge.

In embryonic development, every cell needs to know what it has to become. It would be inappropriate, for instance, for a cell located where the skin will be to become a liver cell. However, a cell does not necessarily know where in the body it is located. To inform each cell of its precise position and eventual fate, the developing embryo relies on complex, overlapping gradients of multiple secreted proteins (patterning factors) that activate molecules on the surface of the cells that, in turn, alter the gene expression patterns of these cells. The history of the previous signals is then recorded in the DNA of the cell by chemical alterations to both the histones and DNA.

Thus, the fate determination of each cell in development depends on a variety of inputs, including the timing of exposure to gradients of patterning factors, the cell's previous developmental history, and types and concentrations of patterning factors that the cell experiences. Interactions with other cells and the local microenvironment also play a considerable role, including determining whether a given cell will survive or die. The cues and responses of the cells can be stunningly complex, and development is incompletely understood even for the best-studied cell types. Even in the cleavage-stage embryo, where the system is quite simple, the signaling pathway that differentiates inside from outside cells was discovered only in 2009 (Nishioka et al. 2009).

In the context of this complexity, it is stunning that we are at all able to differentiate cells along particular pathways. Most stem cell differentiation protocols are far from 100 % efficient when it comes to the phenotype of the cells that they output. When contemplating that we do not actually understand most of the interactions that occur during development, and that most differentiation protocols use cell aggregates, it is quite clear that intercellular signaling within the dish is an important component of stem cell differentiation protocols.

In the hypothalamus, one published protocol indeed relied on self-patterning of mouse ES cells. In brief, cells were allowed to aggregate and develop with as few (known) disruptive chemical cues as possible (Wataya et al. 2008). The success of this protocol suggested that the hypothalamic cell fate is developmentally rather simple and relies on few cues in order to be induced. Cut off from external gradients, the cells produced hypothalamus almost by default. This approach appeared to produce a number of neurons of different types expressing markers found in the hypothalamus, so no particular peptide-secreting cell was the default fate. An alternative explanation is that, in the absence of environmental cues, cells tended towards fates that secreted the molecules necessary to induce the hypothalamus.

However, in human ES cells, this protocol is considerably less efficient, so two directed differentiation protocols have recently been published. This protocol seems to produce a mixture of hypothalamic-like neurons (particularly, neuronal subtypes found in the ventral hypothalamus; Merkle et al. 2015; Wang et al. 2015).

These neuronal mixtures are as close as we have gotten to producing individual subtype hypothalamic-like neurons, and while they are a good start, the complex microenvironment of that brain region creates problems for derivation of more specific cell types. A lack of information about the developmental cues guiding the specification of many hypothalamic cell types compounds this problem.

Direct Reprogramming: An Alternative Pathway to Obtaining Patient-Matched Neuron-Like Cells

The discovery that cells could be induced to acquire an embryonic-like cell fate by treatment with just four viruses to change gene expression naturally led to the question of whether specific types of neurons could be obtained in a similar manner. The answer from the field, thus far, seems to be a resounding "yes." Multiple papers have been published showing that infection of various fully differentiated cell types with viruses is able to produce cells similar to various kinds of neurons (Tsunemoto et al. 2015). However, not all protocols are as simple as the Yamanaka protocol, with some requiring 20+ different genetically modified viruses to enter the same cell in order to be effective. Even with an efficiency of viral delivery of 95 %, such an approach would produce a conversion rate of less than 36 %, assuming every cell infected with virus is converted (which is very unlikely). In practice, the conversion rates are often in the single digits.

The unique challenge of trying to obtain neuronal cells using this method, as opposed to other cell types such as pluripotent cells or hepatocytes, is that neuronal cells do not replicate. Thus, while for most other cell types it is possible to feed them media that will allow replication of large numbers of that cell type at the expense of others, this is not the case in neurons.

An additional concern is that introducing so many different viruses into cells is likely to induce mutations, which could interfere with the normal function of the cells and alter their properties. In addition, these mutations would present a high risk of carcinogenesis when transferred into patients, making neurons obtained in this manner poor candidates for transplantation therapies. It is possible that direct conversion of neurons by other means, such as small molecules or delivery of micro-RNAs, will circumvent both the efficiency and mutagenesis concerns.

Relevance of In Vitro Cell Types to Neuronal Biology

Nearly every discussion of in vitro modeling would have to start with the recognition that in vitro models lack many of the factors found within an intact organism and that many aspects of the conditions found in vitro (for instance, high concentrations of oxygen and a lack of cell-to-cell contact in three dimensions) could interfere with cell survival and function, giving rise to artifacts once cells are studied in culture. It also cannot be denied that certain models in culture reflect aspects of conditions within organisms better than others. Every batch of cells differentiated from stem cells needs to be quality controlled to ascertain whether the cell type being cultured reflects particular aspects of biology within the intact organism.

Since the purpose of differentiating stem cells is fundamentally to make a particular kind of cell normally found within the body, it is important to produce a comprehensive and applicable definition of cell type. This task is complicated by more and more data from single-cell RNAseq and electrophysiology studies that are demonstrating considerable molecular and functional variation within cell populations that would normally be defined as being the same "type." One possibility is to define a cell type as a population of cells within a range of phenotypes that perform analogous functions within the intact organism (reviewed Tabansky et al. 2016). Such a definition would naturally exclude any in vitro cell type, as that cell type is found outside the organism, and it is philosophically impossible to rule out the possibility of an undetected difference between a cell in culture and its counterpart in the organism. Therefore, the aim should not be to faithfully replicate every aspect of an in vivo cell type in culture but, instead, to produce a number of models that reflect the interesting features of a cell type as closely as possible.

This way, each new discovery made with a single model can be subjected to multiple functional tests before being tested again in an organism. Using this strategy, false discovery rates from in vitro models should be decreased.

Outlook

In summary, pluripotent stem cells offer a promising path to understanding and treating neuroendocrine diseases. Considerable challenges remain before we are able to transplant neurons derived from these cells into patients, but studying them in culture might be more accessible. Using induced pluripotent stem cells, we can produce cells that are genetically matched to patients to model development and disease. However, in creating cells that can be used in culture, it is important to keep in mind that it may be impossible to faithfully mimic every aspect of the environment that they encounter in an organism, and thus the cells in culture may behave differently than they would in a brain. It is, therefore, useful to create multiple, redundant models of each cell type, so that false discovery can be minimized. It is likely that the field will continue to advance rapidly, and that it will produce considerable insights for neuroendocrinology.

References

Chung YG, Eum JH, Lee JE, Shim SH, Sepilian V, Hong SW, Lee Y, Treff NR, Choi YH, Kimbrel EA, Dittman RE, Lanza R, Lee DR (2014) Human somatic cell nuclear transfer using adult cells. Cell Stem Cell 14:777–780

Greber B, Wu G, Bernemann C, Joo JY, Han DW, Ko K, Tapia N, Sabour D, Sterneckert J, Tesar P, Schöler HR (2010) Conserved and divergent roles of FGF signaling in mouse epiblast stem cells and human embryonic stem cells. Cell Stem Cell 6:215–226

Marchetto MC, Gage FH (2012) Modeling brain disease in a dish: really? Cell Stem Cell 10:642–645

Merkle FT, Maroof A, Wataya T, Sasai Y, Studer L, Eggan K, Schier AF (2015) Generation of neuropeptidergic hypothalamic neurons from human pluripotent stem cells. Development 142:633–643

Nishioka N, Inoue K, Adachi K, Kiyonari H, Ota M, Ralston A, Yabuta N, Hirahara S, Stephenson RO, Ogonuki N, Makita R, Kurihara H, Morin-Kensicki EM, Nojima H, Rossant J, Nakao K, Niwa H, Sasaki H (2009) The Hippo signaling pathway components Lats and Yap pattern Tead4 activity to distinguish mouse trophectoderm from inner cell mass. Develop Cell 16:398–410

Schlaeger TM, Daheron L, Brickler TR, Entwisle S, Chan K, Cianci A, DeVine A, Ettenger A, Fitzgerald K, Godfrey M, Gupta D, McPherson J, Malwadkar P, Gupta M, Bell B, Doi A, Jung N, Li X, Lynes MS, Brookes E, Cherry ABC, Demirbas D, Tsankov AM, Zon LI, Rubin LL, Feinberg AP, Meissner A, Cowan CA, Daley GQ (2015) A comparison of non-integrating reprogramming methods. Nat Biotech 33:58–63

Schwartz SD, Hubschman JP, Heilwell G, Franco-Cardenas V, Pan CK, Ostrick RM, Mickunas E, Gay R, Klimanskaya I, Lanza R (2012) Embryonic stem cell trials for macular degeneration: a preliminary report. Lancet 379:713–720

Tabansky I, Lenarcic A, Draft RW, Loulier K, Keskin DB, Rosains J, Rivera-Feliciano J, Lichtman JW, Livet J, Stern JN, Sanes JR, Eggan K (2013) Developmental bias in cleavage-stage mouse blastomeres. Curr Biol 23:21–31

Tabansky I, Stern J, Pfaff DW (2016) Front Behav Neurosci. doi:10.3389/fnbeh.2015.00342

Tabar V, Studer L (2014) Pluripotent stem cells in regenerative medicine: challenges and recent progress. Nat Rev Genet 15:82–92

Takahashi K, Yamanaka S (2006) Induction of pluripotent stem cells from mouse embryonic and adult fibroblast cultures by defined factors. Cell 126:663–676

Tesar PJ, Tesar PJ, Chenoweth JG, Brook FA, Davies TJ, Evans EP, Mack DL, Gardner RL, McKay RD (2007) New cell lines from mouse epiblast share defining features with human embryonic stem cells. Nature 448:196–199

Thomson JA (1998) Embryonic stem cell lines derived from human blastocysts. Science 282:1145–1147

Tsunemoto RK, Eade KT, Blanchard JW, Baldwin KK (2015) Forward engineering neuronal diversity using direct reprogramming. EMBO J 34:1445–1455

Wang L, Meece K, Williams DJ, Lo KA, Zimmer M, Heinrich G, Martin Carli J, Leduc CA, Sun L, Zeltser LM, Freeby M, Goland R, Tsang SH, Wardlaw SL, Egli D, Leibel RL (2015) Differentiation of hypothalamic-like neurons from human pluripotent stem cells. J Clin Invest 125:796–808

Wataya T, Ando S, Muguruma K, Ikeda H, Watanabe K, Eiraku M, Kawada M, Takahashi J, Hashimoto N, Sasai Y (2008) Minimization of exogenous signals in ES cell culture induces rostral hypothalamic differentiation. Proc Natl Acad Sci USA 105:11796–11801

Human Pluripotent-Derived Lineages for Repairing Hypopituitarism

Lorenz Studer and Viviane Tabar

Abstract Human pluripotent stem cells (hPSCs) present a potentially unlimited source of specialized cell types for regenerative medicine. Over the last few years there has been rapid progress in realizing this potential by developing protocols to generate disease-relevant cell types in vitro on demand. The approach was particularly successful for the nervous system, where the field is at the verge of human translation for several indications, including the treatment of eye disorders, Parkinson's disease and spinal cord injury. More recently, there has also been success in deriving anterior pituitary lineages from both mouse and human pluripotent stem cells. In vitro-derived pituitary hormone-producing cell types present an attractive source for repair in patients with hypopituitarism. However, several hurdles remain towards realizing this goal. In particular, there is a need to further improve the efficiency and precision with which specific hormone-producing lineages can be derived. Furthermore, it will be important to assess the potential of both ectopic and orthotopic transplantation strategies to achieve meaningful hormone replacement. The ultimate challenge will be repair that moves beyond hormone replacement towards the full functional integration of the grafted cells into the complex regulatory endocrine network controlled by the human pituitary gland.

L. Studer (✉)
Developmental Biology, The Center for Stem Cell Biology, Memorial Sloan-Kettering Cancer Center, New York, NY, USA
e-mail: studerl@MSKCC.ORG

V. Tabar (✉)
Department of Neurosurgery, The Center for Stem Cell Biology, Memorial Sloan-Kettering Cancer Center, New York, NY, USA
e-mail: tabarv@MSKCC.ORG

© The Author(s) 2016
D. Pfaff, Y. Christen (eds.), *Stem Cells in Neuroendocrinology*, Research and Perspectives in Endocrine Interactions, DOI 10.1007/978-3-319-41603-8_3

Derivation of Human Neural Cell Types for Regenerative Medicine

The isolation of human embryonic stem cells (ESCs; Thomson et al. 1998) and the remarkable feat of reprogramming somatic cells back to pluripotency via induced pluripotent stem cell (iPSC) technology (Takahashi et al. 2007; Takahashi and Yamanaka 2006; Yu et al. 2007) have set the stage for a new era of regenerative medicine. Human pluripotent stem cells (hPSCs), a term comprising both human ESCs and iPSCs, are characterized by their potential to differentiate into any cell lineage of the body. For many years, the main challenge in the field has been to capture the broad differentiation potential of hPSCs towards specific cell lineages relevant to modeling and treating human disease. However, there has been considerable progress recently in establishing differentiation protocol for many key lineages such as endoderm-derived insulin-producing pancreatic cells (Pagliuca et al. 2014; Rezania et al. 2014) for the treatment of diabetes or mesoderm-derived cardiac cells for heart repair (Chong et al. 2014). Some of the most dramatic successes, however, have involved ectoderm-derived lineages, in particular retinal and CNS lineages (for review see Tabar and Studer 2014)). In fact, the very first attempts at translating ESC technology towards the treatment of human patients was based on the use of oligodendrocyte precursor-like cells in patients with spinal cord injury (SCI: Alper 2009; Priest et al. 2015). However, SCI patients represent a challenging target for cell therapy, as the primary defect is a problem of connectivity between the brain and spinal cord rather than the loss of a specific cell type. Currently, the most widely pursued clinical target is the transplantation of hPSC-derived retinal pigment epithelial cells (RPEs) in patients with macular degeneration. There are nearly a dozen different RPE-based clinical trials either ongoing or in the planning phase (Kimbrel and Lanza 2015). Initial results using hESC-derived RPEs suggest that the approach can be translated safely into humans (Schwartz et al. 2015).

Beyond eye disorders, there has been particular interest in developing cell-based therapies for the treatment of various neurodegenerative disorders. In the case of Parkinson's disease (PD), several studies demonstrated excellent in vivo survival of hPSC-derived midbrain dopamine (mDA) neurons in mouse, rat and non-human primate hosts (Kirkeby et al. 2012; Kriks et al. 2011). The transplantation of mDA neurons represents an example of replacing a highly specific neuronal subtype and a strategy that is thought to involve functional integration of the grafted cells into the existing neuronal networks. Indeed, a recent study from our group used optogenetics to demonstrate that functional rescue in the PD host animals depended on the continued neuronal activity of the grafted hESC-derived mDA neurons, and "switching-off" the graft led to a reversal of functional benefit within minutes (Steinbeck et al. 2015). The ability to derive mDA neurons from hESCs and hiPSCs and the promising pre-clinical data have set the stage for ongoing translational efforts towards testing this approach in human PD patients. Clinical trials are being planned in the US, Japan and Sweden, which have led to the formation of G-Force

PD, a global effort to coordinate hPSC-based cell therapy efforts in PD (Barker et al. 2015). Another neurodegenerative disease being targeted is Huntington's disease where several protocols have been published to generate authentic, striatal medium spiny neurons and where there is some initial evidence of efficacy in preclinical models (Arber et al. 2015; Delli Carri et al. 2013; Ma et al. 2012). Finally, several promising strategies are under development using glial cells. These include the transplantation of hPSC-derived oligodendrocytes in genetic models of white matter loss (Wang et al. 2013) and the remyelination of the brain following radiation-induced brain damage (Piao et al. 2015), a common and serious problem in cancer patients subjected to cranial irradiation (Greene-Schloesser et al. 2012; Schatz et al. 2000).

With our increasing ability to generate potentially any neural lineage on demand, the main challenge in the field has moved beyond making a specific cell type towards translation and therapy development in regenerative medicine. While the initial therapeutic targets for cell therapy are focused on replacing highly defined populations of cells such as RPEs or mDA neurons, it may be necessary in future studies to replace multiple cell types in combination to achieve meaningful rescue in a broader range of human disorders. A particular challenge for neuronal cell therapies is the importance of developing pre-clinical and ultimately clinical evidence that in vitro-derived cells can integrate into the complex circuitry of the human brain.

Derivation and Application of Human Pituitary Lineages

Replacing endocrine cells is conceptually more straightforward than replacing CNS neurons because there is no need to re-establish a complex synaptic circuitry to achieve improved function. However, the pituitary gland is also highly complex and acts as the master regulator of endocrine function, controlling a diverse range of responses in the body including stress control, growth and sexual function. Such complexity makes any treatment of hypopituitarism - the loss of pituitary function – challenging, as many hormones need to be replaced in a coordinated manner. In the context of cell therapy, this requires the ability to generate multiple hormone-producing cells at scale and on demand. To date, the main focus of hPSC-based approaches for treating endocrine disorders has been on the treatment of type I diabetes (Bruin et al. 2015). One key rationale for proposing a cell-based approach in diabetes is successful derivation of functional islet cells from hPSCs (Pagliuca et al. 2014; Rezania et al. 2014) and the expectation that grafted pancreatic β-cells will establish a feedback loop sensing glucose and adjusting insulin levels continuously throughout the day, something that is difficult to achieve by insulin injections. Furthermore, it appears likely that regulatory control can be achieved with cells that are not placed orthotopically into the pancreas but injected into a surgically more accessible tissue with high vascularity, such as the spleen or liver (Bruin et al. 2015). In contrast, orthotopic placement may be more critical for pituitary cells that respond to rapid and short acting signals from the hypothalamus. The challenge of recreating anterior pituitary lineage diversity in vitro was first tackled using mouse ESC cells. In a seminal study by the Sasai lab (Suga et al. 2011), a differentiation protocol was presented that allowed the derivation of mouse pituitary lineages via a serum-free embryoid body (SFEBq) culture step. SFEBq conditions were initially developed to generate forebrain lineages from mouse (Watanabe et al. 2005) and subsequently from human ESCs (Eiraku et al. 2008). In contrast to the forebrain, which is derived from the CNS, the anterior pituitary gland is derived from the oral ectoderm, which is part of the cranial placode lineages during development. To direct cell lineage towards oral ectoderm, Suga et al. (2011) showed that BMP4 exposure could trigger the induction of PITX2, an oral ectoderm marker, at the periphery of the differentiating SFEBs. Subsequent exposure to agonists of sonic hedgehog (SHH) signaling triggered expression of LHX3, which is a definitive anterior pituitary lineage marker. One remarkable feature during the induction process is the morphogenetic movements of the oral ectoderm that mimic the formation of Rathke's pouch, an invagination of the oral ectoderm occurring during development that results in anterior pituitary gland formation. However, the overall efficiency of generating Pitx2+ oral ecto-derm cells was low and most cells in the SFEBq cultures retained a neural identity. Approximately 1–7 % of the non-neural cells expressed specific hormones, a number that was dependent on further modulation of WNT activation for induction of growth hormone- (GH) or prolactin (PRL)-producing cells or inhibition of Notch signaling for obtaining ACTH+ cells (Suga et al. 2011). For the in vitro-derived ACTH+ cells, Suga et al. (2011) demonstrated CRH-dependent hormone secretion in vitro. Furthermore, transplanted cells were able to survive in vivo in an animal with surgically induced hypopituitarism, and they extended the life span of those animals, presumably by partially restoring their stress response. Some of the key questions raised by the Suga et al. study include whether the same technology can be applied for human ESC and iPSCs, whether the overall yield of anterior pituitary placode and hormone-producing cells can be improved and whether the 3D culture step, allowing the interaction of oral ectoderm-like and hypothalamic tissue, is critical for the efficient induction of anterior pituitary lineage cells.

Some initial answers to these questions came from an independent effort in our laboratories aimed at inducing cranial placode lineages from hPSCs (Dincer et al. 2013). Similar to the SFEBq technology, the human placode induction strategy was based on a protocol, dual-SMAD inhibition (Chambers et al. 2009), that was initially developed for inducing forebrain fates. Under dual-SMAD inhi-bition conditions, a monolayer of human ESCs or iPSCs can be converted at nearly 100 % efficiency into PAX6+ anterior neuroectoderm within about 10 days of differentiation (Chambers et al. 2009). Dual-SMAD inhibition involves concomi-tant exposure of hPSCs to inhibitors of BMP signaling (either Noggin or the ALK2/3 inhibitor LDN193189) and inhibitors of TGFβ, Activin and Nodal signaling (commonly via the small molecule compound SB431542). In Dincer et al. (2013), we showed that the key difference between CNS versus placode induction was the inhibition versus activation of BMP signaling. In contrast to the induction of CNS lineage under conditions of dual-SMAD inhibition, placode induction requires the timed removal of the BMP inhibitor at 48 h after neural induction, allowing endogenous BMP signaling to rebound. Under these default placode induction conditions, the majority of the hPSC-derived cells expressed PAX3, suggestive of trigeminal placode fate (Dincer et al. 2013). However, upon activation of SHH signaling, there was a marked increase in oral ectoderm markers such as PITX1 and SIX6. Further differentiation of these pituitary placode precursors was shown to yield various hormone-expressing cells, including ACTH+, FSH+ and GH+ lineages with clear evidence of in vitro hormone release. Finally, our study demonstrated that subcutaneous injection of hPSC-derived pituitary precursors into nude rats yielded measurable levels of ACTH and GH secretion in vivo. The findings of Dincer et al. suggested that the robust induction of human pituitary hormone-expressing cells did not require a 3D culture step. However, the overall efficiency of pituitary placode induction remained suboptimal. Furthermore, the media conditions during differentiation included components such as knockout serum-replacement (KSR) that are known to introduce batch-to-batch variability into the differentiation process. Furthermore, the study did not attempt to enrich for specific hormone lineages that may be required to develop better tailored therapies for each individual patient. A first step towards optimizing cranial placode induction in the absence of KSR was achieved by another study that carefully optimized the timing of BMP4 application (Leung et al. 2013). The results indicated that early exposure to BMP4 could increase overall cranial placode yield whereas the subsequent inhibition versus activation of BMP at later stages of differentiation could modulate the regional identity of hESC-derived placodal cells from PAX6+ anterior to PAX3+ posterior placode (Leung et al. 2013). Finally, the study confirmed that activation of SHH signaling increased the expression of oral ectoderm markers including *PITX1* and *PITX2*.

More recently, members of the Sasai lab presented a study that adapted their 3D approach to human cells (Ozone et al. 2016). The study was based on a modified SFEBq culture system triggering differentiation in the presence of KSR, SHH agonist and BMP4 to yield 3D structures composed of hypothalamic cells in the center of the 3D aggregates and oral ectoderm cells at the periphery. In a proportion of those structures, the authors again observed the spontaneous formation of Rathke's pouch-like structures similar to their original data in mouse ESCs, though at a lower frequency. While the overall induction efficiency of definitive pituitary lineages remained low, the authors were able to demonstrate both basal and CRH-induced release of ACTH in vitro that was shown to be suppressed by hydrocortisone treatment. Similarly, induction of GH could be modulated both positively and negatively by exposure to GHRH or somatostatin, respectively. Finally, subcutaneous injection of the 3D aggregates into mice with surgical hypophysectomy showed evidence of in vivo ACTH production that was responsive to CRH treatment. The transplanted cells also triggered significant, albeit very low, levels of corticosterone production and led to improved body weight and survival as compared to sham-grafted animals in hypophysectomized hosts. Some of the key remaining challenges include further improving the efficiency and reliability of pituitary lineage differentiation with a greater percentage of hormone-producing cells.

Perspectives and Challenges on the Road to Translation

The adenohypophysis (pituitary) is a remarkable endocrine organ that orchestrates the function of multiple targets via secretion of a set of regulatory hormones in charge of vital functions such as development, growth, puberty, reproduction, lactation and, crucially, response to stress. It receives regulatory endocrine input from the adjacent hypothalamus through a portal circulation system and communicates with the rest of the organism via an extensive network of vessels. Multi-tiered feedback is integrated by this master gland, leading to hormonal and metabolic homeostasis (Tabar 2011). The role of regenerative approaches to the adenohypophysis has received very little attention despite the prevalence of pituitary disorders and the large number of patients requiring pituitary hormone replacement due to traumatic brain injury, genetic, sporadic or iatrogenic disease. Several syndromes of pituitary deficiencies are recognized in humans as the result of mutations of early transcription factors or cell cycle regulator proteins (Melmed 2011). One of the prevalent causes of pituitary deficiency is post-treatment pituitary and hypothalamic damage. Specifically, and of interest to us, is a group of patients who suffer from hypopituitarism as a consequence of brain radiation for a variety of disorders, including hematological malignancies, head and neck cancers, brain tumors and sellar lesions (Appelman-Dijkstra et al. 2011). In fact, growing interest in cancer survivorship has identified hypopituitarism as a major contributor to poor quality of life indices (Darzy 2009). The clinical consequences are extensive and include fatigue, poor concentration, decreased memory and general cognitive abilities as well as significantly reduced well-being. In children these consequences are compounded by more serious learning difficulties and growth and skeletal problems, as well as a major impact on puberty and sexual function (Chemaitilly and Sklar 2010). Radiation damage to the hypothalamus and pituitary regions is progressive and irreversible. Current treatment consists of life-long multiple hormone replacement therapies, a suboptimal solution since static delivery of these molecules is a poor substitute for normal pituitary gland features such as the dynamic secretion of hormones in response to circadian patterns, feedback mechanisms or stressful conditions. In addition, treatment can be prohibitively expensive, with costs of growth hormone replacement alone exceeding $20,000 per year.

One of the key challenges to restorative strategies, regardless of the etiology of hypopituitarism, involves decisions regarding orthotopic or ectopic graft implantation. Grafts of pituitary tissue or primary cell suspensions from human fetal or rodent sources have been performed extensively using ectopic (Fu and Vankelecom 2012) or more or less orthotopic placement in the pituitary (Falconi and Rossi 1964), hypothalamus (Tulipan et al. 1985) or in the third ventricle (Vuillez et al. 1989). Overall, pituitary tissue or cells survive very well with the exception of conditions of immunological mismatch. A major concern with ectopic (i.e., subcutaneous or kidney) placement is the absence of hypothalamic control. The adenohypophysis is connected to the hypothalamus by a portal vein system that allows the immediate delivery of hypothalamic factors, thus bypassing the systemic circulation. Data from transplants in the hypothalamus, third ventricle or the hypophysis sites suggest that pituitary grafts demonstrate improved function and response to feedback when they are in immediate contact with the hypothalamus (Harris and Jacobsohn 1952; Maxwell et al. 1998). Some of the most successful results have been obtained upon transplantation in hypophysectomized female rats at the level of the median eminence, with good outcomes including restoration of estrus cycles, ability to conceive and lactate pups, adequate growth hormone and ACTH levels, as well as near normalization of body size in comparison to normal controls (Harris and Jacobsohn 1952). Data from transplantation of mouse ESC-derived pituitary cells (injected in the renal capsule) suggested statistically significant elevation of basal ACTH and corticosterone (Suga et al. 2011). From a translational perspective, ectopic placement in the subcutaneous tissue offers significant advantages due to the low risk of the intervention (e.g., in a subcutaneous location) and easy accessibility in case of complications. The experimental evaluation of ectopic grafts may therefore be a justifiable strategy, though integration within the hypothalamic-pituitary-target organ axis and homeostatic control is more likely to be achieved if grafts are placed in the vicinity of the hypothalamus or even within the gland itself. Interestingly, in humans, placement in the sella is simpler than in rodents, due to the development of minimal invasive transnasal endoscopic approaches to the sella and anterior skull base. Experimental evidence of hypothalamic control upon grafting of pituitary cells requires complex assays and readouts, including stimulation tests [e.g., response to thyrotropin releasing hormone (TSH) or to growth hormone-releasing hormone (GHRH) etc.], physiological stress tests [e.g., exposure to cold, arginine testing followed by evaluation of variations in pituitary hormone levels (Akalan et al. 1988; Fisker et al. 1999; Guillemin 2005)]. Behavioral testing can also contribute to the evaluation of the integrity of the hypothalamic-pituitary-target organ axis and its feedback loops. Demonstrating appropriate integration into the neuroendocrine system and its physiological and homeostatic feedback loops should be considered an important component of both efficacy and safety of this strategy. Obviously uncontrolled or random secretion of key hormones such as ACTH or growth hormone can have serious negative health consequences.

Additional considerations in translational strategies would include the possibility of grafting specific pituitary sublineages, e.g., ACTH- or growth hormone-secreting. This approach might require more sophisticated differentiation protocols and, likely, selection strategies that should be compatible with good manufacturing practice (GMP) conditions and safety standards.

Conclusion

The recent successful derivation of pituitary placode lineage and the range of anterior pituitary hormone-producing cells are very exciting advances that will likely herald the development of restorative strategies in humans. Several challenges along the road to translation remain to be tackled. Key questions include the ability to develop selective pituitary sublineages that produce a single target hormone, the development of grafting strategies for human patients, and the demonstration of integration of the grafted cells into the hypothalamic-pituitary-peripheral target axis, a goal that is fundamental to the safety assessment of the cell therapy-based approach to hypopituitarism.

Note Added in Proof
A recent manuscript from our team presents a fully defined and efficient protocol for the derivation of anterior pituitary hormone producing cells from human pluripotent stem cells and demonstrates hormone release in a rat model of hypopituitarism (Zimm et al. 2016).

References

Akalan N, Pamir MN, Benli K, Erbengi A, Erbengi T (1988) Fetal pituitary transplants into the hypothalamic area of hypophysectomized rats. Surg Neurol 30:342–349
Alper J (2009) Geron gets green light for human trial of ES cell-derived product. Nat Biotechnol 27:213–214
Appelman-Dijkstra NM, Kokshoorn NE, Dekkers OM, Neelis KJ, Biermasz NR, Romijn JA, Smit JW, Pereira AM (2011) Pituitary dysfunction in adult patients after cranial radiotherapy: systematic review and meta-analysis. J Clin Endocrinol Metab 96:2330–2340
Arber C, Precious SV, Cambray S, Risner-Janiczek JR, Kelly C, Noakes Z, Fjodorova M, Heuer A, Ungless MA, Rodriguez TA, Rosser AE, Dunnett SB, Li M (2015) Activin A directs striatal projection neuron differentiation of human pluripotent stem cells. Development 142: 1375–1386
Barker RA, Studer L, Cattaneo E, Takahashi J (2015) G-Force PD: a global initiative in coordinating stem cell-based dopamine treatments for Parkinson's disease. Npj Parkinson's Dis 15017. doi:10.1038/npjparkd.2015.17
Bruin JE, Rezania A, Kieffer TJ (2015) Replacing and safeguarding pancreatic beta cells for diabetes. Sci Translat Med 316ps323
Chambers SM, Fasano CA, Papapetrou EP, Tomishima M, Sadelain M, Studer L (2009) Highly efficient neural conversion of human ES and iPS cells by dual inhibition of SMAD signaling. Nat Biotechnol 27:275–280
Chemaitilly W, Sklar CA (2010) Endocrine complications in long-term survivors of childhood cancers. Endocr Relat Cancer 17:R141–159
Chong JJ, Yang X, Don CW, Minami E, Liu YW, Weyers JJ, Mahoney WM, Van Biber B, Cook SM, Palpant NJ, Gantz JA, Fugate JA, Muskheli V, Gough GM, Vogel KW, Astley CA, Hotchkiss CE, Baldessari A, Pabon L, Reinecke H, Gill EA, Nelson V, Kiem HP, Laflamme MA, Murry CE (2014) Human embryonic-stem-cell-derived cardiomyocytes regenerate non-human primate hearts. Nature 510:273–277

Darzy KH (2009) Radiation-induced hypopituitarism after cancer therapy: who, how and when to test. Nat Clin Pract Endocrinol Metab 5:88–99

Delli Carri A, Onorati M, Lelos MJ, Castiglioni V, Faedo A, Menon R, Camnasio S, Vuono R, Spaiardi P, Talpo F, Toselli M, Martino G, Barker RA, Dunnett SB, Biella G, Cattaneo E (2013) Developmentally coordinated extrinsic signals drive human pluripotent stem cell differentiation toward authentic DARPP-32+ medium-sized spiny neurons. Development 140: 301–312

Dincer Z, Piao J, Niu L, Ganat Y, Kriks S, Zimmer B, Shi SH, Tabar V, Studer L (2013) Specification of functional cranial placode derivatives from human pluripotent stem cells. Cell Rep 5:1387–1402

Eiraku M, Watanabe K, Matsuo-Takasaki M, Kawada M, Yonemura S, Matsumura M, Wataya T, Nishiyama A, Muguruma K, Sasai Y (2008) Self-organized formation of polarized cortical tissues from ESCs and its active manipulation by extrinsic signals. Cell Stem Cell 3:519–532

Falconi G, Rossi GL (1964) Method for placing a pituitary graft into the evacuated pituitary capsule of the hypophysectomized rat or mouse. Endocrinology 75:964–967

Fisker S, Nielsen S, Ebdrup L, Bech JN, Christiansen JS, Pedersen B, Jorgensen JO (1999) The role of nitric oxide in L-arginine-stimulated growth hormone release. J Endocrinol Invest 22: 89–93

Fu Q, Vankelecom H (2012) Regenerative capacity of the adult pituitary: multiple mechanisms of lactotrope restoration after transgenic ablation. Stem Cells Dev 21:3245–3257

Greene-Schloesser D, Robbins ME, Peiffer AM, Shaw EG, Wheeler KT, Chan MD (2012) Radiation-induced brain injury: a review. Front Oncol 2:73

Guillemin R (2005) Hypothalamic hormones a.k.a. hypothalamic releasing factors. J Endocrinol 184:11–28

Harris GW, Jacobsohn D (1952) Functional grafts of the anterior pituitary gland. Proc R Soc Lond Ser B Biol Sci 139:263–276

Kimbrel EA, Lanza R (2015) Current status of pluripotent stem cells: moving the first therapies to the clinic. Nat Rev Drug Discov 14:681–692

Kirkeby A, Grealish S, Wolf DA, Nelander J, Wood J, Lundblad M, Lindvall O, Parmar M (2012) Generation of regionally specified neural progenitors and functional neurons from human embryonic stem cells under defined conditions. Cell Rep 1:703–714

Kriks S, Shim JW, Piao J, Ganat YM, Wakeman DR, Xie Z, Carrillo-Reid L, Auyeung G, Antonacci C, Buch A, Yang L, Beal MF, Surmeier DJ, Kordower JH, Tabar V, Studer L (2011) Dopamine neurons derived from human ES cells efficiently engraft in animal models of Parkinson's disease. Nature 480:547–551

Leung AW, Kent Morest D, Li JY (2013) Differential BMP signaling controls formation and differentiation of multipotent preplacodal ectoderm progenitors from human embryonic stem cells. Dev Biol 379:208–220

Ma L, Hu B, Liu Y, Vermilyea SC, Liu H, Gao L, Sun Y, Zhang X, Zhang SC (2012) Human embryonic stem cell-derived GABA neurons correct locomotion deficits in quinolinic acid-lesioned mice. Cell Stem Cell 10:455–464

Maxwell M, Allegra C, MacGillivray J, Hsu DW, Hedley-Whyte ET, Riskind P, Madsen JR, Black PM (1998) Functional transplantation of the rat pituitary gland. Neurosurgery 43:1157–1163

Melmed S (2011) Pathogenesis of pituitary tumors. Nat Rev Endocrinol 7:257–266

Ozone C, Suga H, Eiraku M, Kadoshima T, Yonemura S, Takata N, Oiso Y, Tsuji T, Sasai Y (2016) Functional anterior pituitary generated in self-organizing culture of human embryonic stem cells. Nat Commun 7:10351

Pagliuca FW, Millman JR, Gurtler M, Segel M, Van Dervort A, Ryu JH, Peterson QP, Greiner D, Melton DA (2014) Generation of functional human pancreatic beta cells in vitro. Cell 159: 428–439

Piao J, Major T, Auyeung G, Policarpio E, Menon J, Droms L, Gutin P, Uryu K, Tchieu J, Soulet D, Tabar V (2015) Human embryonic stem cell-derived oligodendrocyte progenitors remyelinate the brain and rescue behavioral deficits following radiation. Cell Stem Cell 16: 198–210

Priest CA, Manley NC, Denham J, Wirth ED 3rd, Lebkowski JS (2015) Preclinical safety of human embryonic stem cell-derived oligodendrocyte progenitors supporting clinical trials in spinal cord injury. Regen Med 10:939–958

Rezania A, Bruin JE, Arora P, Rubin A, Batushansky I, Asadi A, O'Dwyer S, Quiskamp N, Mojibian M, Albrecht T, Yang YH, Johnson JD, Kieffer TJ (2014) Reversal of diabetes with insulin-producing cells derived in vitro from human pluripotent stem cells. Nat Biotechnol 32:1121–1133

Schatz J, Kramer JH, Ablin A, Matthay KK (2000) Processing speed, working memory, and IQ: a developmental model of cognitive deficits following cranial radiation therapy. Neuropsychology 14:189–200

Schwartz SD, Regillo CD, Lam BL, Eliott D, Rosenfeld PJ, Gregori NZ, Hubschman JP, Davis JL, Heilwell G, Spirn M, Maguire J, Gay R, Bateman J, Ostrick RM, Morris D, Vincent M, Anglade E, Del Priore LV, Lanza R (2015) Human embryonic stem cell-derived retinal pigment epithelium in patients with age-related macular degeneration and Stargardt's macular dystrophy: follow-up of two open-label phase 1/2 studies. Lancet 385:509–516

Steinbeck JA, Choi SJ, Mrejeru A, Ganat Y, Deisseroth K, Sulzer D, Mosharov EV, Studer L (2015) Optogenetics enables functional analysis of human embryonic stem cell-derived grafts in a Parkinson's disease model. Nat Biotechnol 33:204–209

Suga H, Kadoshima T, Minaguchi M, Ohgushi M, Soen M, Nakano T, Takata N, Wataya T, Muguruma K, Miyoshi H, Yonemura S, Oiso Y, Sasai Y (2011) Self-formation of functional adenohypophysis in three-dimensional culture. Nature 480:57–62

Tabar V (2011) Making a pituitary gland in a dish. Cell Stem Cell 9:490–491

Tabar V, Studer L (2014) Pluripotent stem cells in regenerative medicine: challenges and recent progress. Nat Rev Genet 15:82–92

Takahashi K, Yamanaka S (2006) Induction of pluripotent stem cells from mouse embryonic and adult fibroblast cultures by defined factors. Cell 126:663–676

Takahashi K, Tanabe K, Ohnuki M, Narita M, Ichisaka T, Tomoda K, Yamanaka S (2007) Induction of pluripotent stem cells from adult human fibroblasts by defined factors. Cell 131: 861–872

Thomson JA, Itskovitz-Eldor J, Shapiro SS, Waknitz MA, Swiergiel JJ, Marshall VS, Jones JM (1998) Embryonic stem cell lines derived from human blastocysts. Science 282:1145–1147

Tulipan NB, Zacur HA, Allen GS (1985) Pituitary transplantation: Part 1. Successful reconstitution of pituitary-dependent hormone levels. Neurosurgery 16:331–335

Vuillez P, Moos F, Stoeckel ME (1989) Immunocytochemical and ultrastructural studies on allografts of the pituitary neurointermediate lobe in the third cerebral ventricle of the rat. Cell Tissue Res 255:393–404

Wang S, Bates J, Li X, Schanz S, Chandler-Militello D, Levine C, Maherali N, Studer L, Hochedlinger K, Windrem M, Goldman SA (2013) Human iPSC-derived oligodendrocyte progenitor cells can myelinate and rescue a mouse model of congenital hypomyelination. Cell Stem Cell 12:252–264

Watanabe K, Kamiya D, Nishiyama A, Katayama T, Nozaki S, Kawasaki H, Watanabe Y, Mizuseki K, Sasai Y (2005) Directed differentiation of telencephalic precursors from embryonic stem cells. Nat Neurosci 8:288–296

Yu J, Vodyanik MA, Smuga-Otto K, Antosiewicz-Bourget J, Frane JL, Tian S, Nie J, Jonsdottir GA, Ruotti V, Stewart R, Slukvin II, Thomson JA (2007) Induced pluripotent stem cell lines derived from human somatic cells. Science 318:1917–1920

Zimmer B, Piao J, Ramnarine K, Tomishima MJ, Tabar V, Studer L (2016) Derivation of diverse hormone-releasing pituitary cells from human pluripotent stem cells. Stem Cell Rep 6(6):858–872. doi:10.1016/j.stemcr.2016.05.005, PMID: 27304916

Recapitulating Hypothalamus and Pituitary Development Using Embryonic Stem/ Induced Pluripotent Stem Cells

Hidetaka Suga

Abstract The hypothalamic-pituitary system is essential for maintaining life and controlling systemic homeostasis. However, it can be negatively affected by various diseases, resulting in life-long serious symptoms.

Pluripotent stem cells, such as embryonic stem (ES) cells and induced pluripotent stem (iPS) cells, differentiate into neuroectodermal progenitors when cultured as floating aggregates under serum-free conditions.

Recent results have shown that strict removal of exogenous patterning factors during the early differentiation period induces efficient generation of rostral hypothalamic-like progenitors from mouse ES cell-derived neuroectodermal cells. The use of growth factor-free, chemically defined medium was critical for this induction. The ES cell-derived hypothalamic-like progenitors generated rostral-dorsal hypothalamic neurons, in particular magnocellular vasopressinergic neurons, which release hormones upon stimulation.

We subsequently reported efficient self-formation of three-dimensional adeno-hypophysis tissues in aggregate cultures of mouse ES cells. The ES cells were stimulated to differentiate into non-neural head ectoderm and hypothalamic neuroectoderm in adjacent layers within the aggregate, followed by treatment with a Sonic Hedgehog agonist. Self-organization of Rathke's pouch-like structures occurred at the interface of the two epithelia in vivo, and various endocrine cells, including corticotrophs and somatotrophs, were subsequently produced. The corticotrophs efficiently secreted adrenocorticotropic hormone in response to corticotropin-releasing hormone. Furthermore, when engrafted in vivo, these cells rescued systemic glucocorticoid levels in hypopituitary mice.

The present study aimed to prepare hypothalamic and pituitary tissues from human pluripotent stem cells and establish effective transplantation techniques for future clinical applications. Preliminary results indicated differentiation using human ES/iPS cells, and the culture method replicated stepwise embryonic differentiation. Therefore, these methods could potentially be used as developmental and disease models as well as for future regenerative medicine.

H. Suga (✉)
Department of Endocrinology and Diabetes, Nagoya University Hospital, 65 Tsurumai-cho,
Showa-ku, Nagoya, Aich 466-8550, Japan
e-mail: sugahide@med.nagoya-u.ac.jp

© The Author(s) 2016
D. Pfaff, Y. Christen (eds.), *Stem Cells in Neuroendocrinology*, Research and
Perspectives in Endocrine Interactions, DOI 10.1007/978-3-319-41603-8_4

Introduction

The hypothalamus and adenohypophysis maintain physiological homeostasis by controlling the endocrine system. A collection of studies exploring their development and function has shown that they are essential for the regulation of vital functions. However, their regeneration remains largely unclear.

Recently, somatic stem cells have been recognized as a major source for tissue maintenance and regeneration. A 2005 study reported that somatic stem cells exist in the adenohypophysis (Chen et al. 2005). Subsequent studies have discussed their functions during early postnatal pituitary maturation (Fauquier et al. 2001; Kikuchi et al. 2007; Chen et al. 2009; Gremeaux et al. 2012; Mollard et al. 2012), after pituitary damage (Luque et al. 2011; Fu et al. 2012; Langlais et al. 2013), and in pituitary tumorigenesis (Gaston-Massuet et al. 2011; Andoniadou et al. 2012; Garcia-Lavandeira et al. 2012; Li et al. 2012).

In addition to somatic stem cells, studies have focused on embryonic stem (ES) cells and induced pluripotent stem (iPS) cells. These pluripotent stem cells exhibit self-renewal properties and pluripotent differentiation. Therefore, they have attracted attention as a cell source for differentiated tissues in clinical applications.

A Need for Hypothalamus and Adenohypophysis Regenerative Medicine

The hypothalamus and adenohypophysis are located in adjacent regions, and they coordinate functions as the center for the endocrine systems. Following dysfunction, patients experience various systemic symptoms. Current treatment primarily consists of hormone replacement therapy, but various factors can complicate the proper dose. Drug administration cannot precisely mimic the circadian or stress-induced changes of hormone requirements. For example, we have reported that some patients with central diabetes insipidus show unstable serum Na levels, resulting in a poor prognosis (Arima et al. 2014). This instability is caused by the lack of positive and negative control systems, which is characteristic of hormone-producing cells. As for hypopituitarism, it has been reported that adrenal crisis occurs in a substantial proportion of society, and adrenal crisis-associated mortality is not negligible, even in educated patients (Hahner et al. 2015). Furthermore, adrenocorticotropic hormone (ACTH)-dependent adrenal insufficiency, as well as high-dose hydrocortisone treatment, serves as a predictor for acromegaly-associated mortality (Sherlock et al. 2009; Ben-Shlomo 2010). Taken together, these factors indicate that there are many prospects for pituitary regenerative medicine.

Mouse ES Cells

The establishment of mouse pluripotent ES cells significantly contributed to the advancement of biology and medicine. In 1981, Evans and Kaufman successfully established mouse ES cells from the inner cell mass of mouse blastocyst-stage embryos (Evans and Kaufman 1981). Various knock-out and knock-in mice have been established using genetically modified ES cells, which have contributed to our understanding of gene functions (Bernstein and Breitman 1989; Babinet and Cohen-Tannoudji 2001).

There are two reasons for the use of mouse ES cells, rather than human ES cells, in our recent studies. One reason is the short developmental period; the duration of mouse fetal development is about 20 days, which is much shorter than the 300 days of human development. Therefore, mouse ES cells are suitable for establishing novel differentiation methods with numerous trial-and-error processes. Another reason is the similarity between mouse and human cells. For example, the retinal differentiation method from human ES cells (Nakano et al. 2012) was established based on a previous report using mouse ES cells (Eiraku et al. 2011). Although the inducing culture methods differ considerably, their key principles are similar. The fundamental processes of mouse ES cells appear to be applicable to human ES cells.

Pituitary Gland Embryology

The adenohypophysis, which corresponds to the anterior pituitary gland, contains several types of endocrine cells that secrete factors including adrenocorticotropic hormone (ACTH), growth hormone (GH), prolactin (PRL), thyroid-stimulating hormone, luteinizing hormone, and follicle-stimulating hormone. The posterior pituitary gland consists of axons and terminals of hypothalamic neurons, i.e., vasopressin and oxytocin neurons. The development of the adenohypophysis is a complex process. During early development, the adenohypophysis anlage originates as a placode in the non-neural ectoderm adjacent to the anterior neural plate (Fig. 1a). Both the adenohypophysis placode and hypothalamic anlage interact with each other. Accordingly, the thickened placode invaginates and subsequently detaches from the oral ectoderm to form a hollowed vesicle, termed "Rathke's pouch" (Zhu et al. 2007; Fig. 1b). The molecular nature of this local inductive interaction during this initial phase of pituitary formation remains elusive, but FGF and BMP signals appear to be involved (Takuma et al. 1998; Brinkmeier et al. 2007).

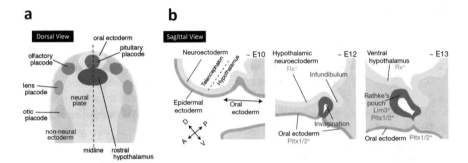

Fig. 1 Diagram of mouse pituitary development. (a) Dorsal view of neural plate and placodes. (b) Sagittal view of pituitary embryogenesis. *E* embryonic day

Three-Dimensional ES Cell Culture

Organ formation during embryogenesis consists of complicated processes that involve various local interactions between different tissues or cells. Despite this complexity, organogenesis can be modeled in vivo. Our colleagues established a three-dimensional culture method for ES cells called "serum-free culture of embryoid body-like aggregates with quick re-aggregation (SFEBq)" (Watanabe et al. 2005; Eiraku et al. 2008). The culture method is quite simple. First, the quality of maintenance for undifferentiated ES cells is very important. For SFEBq culture, maintained ES cells are dissociated to single cells in trypsin or something similar. The cells are then quickly aggregated using low-cell adhesion 96-well plates in differentiation medium suitable for each differentiation purpose.

This culture method is suitable for induction of various ectodermal derivatives from ES cells. In SEFBq cultures, the ES cell aggregates exhibit self-organization (Sasai et al. 2012) and spontaneous formation of a highly ordered structure or patterning. This floating culture has revealed intrinsic programs that drive locally autonomous modes of organogenesis and homeostasis. Using the SFEBq method, mesencephalic dopamine neurons (Kawasaki et al. 2002; Morizane et al. 2006), cortex neurons (Eiraku et al. 2008; Danjo et al. 2011; Kadoshima et al. 2013), the optic cup (Eiraku et al. 2011; Ikeda et al. 2005; Osakada et al. 2008), cerebellar neurons (Muguruma et al. 2010), and hippocampal neurons (Sakaguchi et al. 2015) have been generated from mouse and human ES cells.

Induction of Hypothalamic Neurons from Mouse ES Cells

Using SFEBq cultures, hypothalamic neurons such as vasopressin-positive neurons have been induced from mouse ES cells (Wataya et al. 2008). The differentiation occurs efficiently when the ES cell aggregates are cultured in growth factor-free, chemically defined medium (gfCDM). Strict removal of exogenous patterning factors during early differentiation steps induces efficient generation of rostral hypothalamic-like progenitors (Rax(+)/Six3(+)/Vax1(+); these combinations are characteristic for hypothalamic precursors) in mouse ES cell aggregates. The use of gfCDM is critical. For example, even the presence of exogenous insulin, which is commonly used in cell culture, strongly inhibits differentiation via the Akt-dependent pathway. The ES cell-derived hypothalamic progenitors generate Otp(+)/Brn2(+) neuronal precursors (characteristic of rostral-dorsal hypothalamic neurons) and subsequent magnocellular vasopressinergic neurons that release vasopressin upon stimulation. Additionally, differentiation markers of rostral-"ventral" hypothalamic precursors and neurons have been induced from ES cell-derived Rax (+) progenitors by treatment with Sonic Hedgehog (Shh).

Thus, in the absence of exogenous growth factors in the medium, ES cell-derived neuroectodermal cells spontaneously differentiated into rostral (particularly rostral-dorsal) hypothalamic-like progenitors, which generated characteristic hypothalamic neuroendocrine neurons in a stepwise fashion, as observed in vivo. These findings indicated that, instead of the addition of inductive signals, minimization of exogenous patterning signaling played a key role in rostral hypothalamic specification of neural progenitors derived from pluripotent cells. This work also showed that the default fate of mouse ES cells is the rostral hypothalamus (Wataya et al. 2008).

Two-Layer Formation In Vitro Is the First Step of Adenohypohysis Differentiation

We next established an in vitro differentiation method for the anterior pituitary (38). Rathke's pouch is formed as a result of interactions between the hypothalamus and neighboring oral ectoderm (Zhu et al. 2007). To recapitulate embryonic pituitary development, we co-induced these two tissues within one ES cell aggregate.

Previous results have shown hypothalamic differentiation from mouse ES cells (37). Mouse ES cells can be induced to differentiate into hypothalamic cells when cultured as floating aggregates using the SFEBq method with gfCDM. Therefore, the present study used a technical modification to co-induce oral ectodermal differentiation in addition to hypothalamic differentiation.

We attempted to slightly shift positional information so that the oral ectoderm co-existed with hypothalamic tissues (Suga et al. 2011). As shown in Fig. 1a, the oral ectoderm is generated from the rostral and midline region adjacent to the hypothalamic region in the mouse embryo. Therefore, the rostral and midline shifting information was relevant for mouse ES cell aggregates in the SFEBq culture. We tested many culture conditions known to affect early ectodermal patterning. We ultimately identified two conditions that efficiently induced oral ectoderm. One condition was the addition of bone morphogenetic protein 4 (BMP4). However, treatment with 0.5 µM BMP4 strongly inhibited hypothalamic

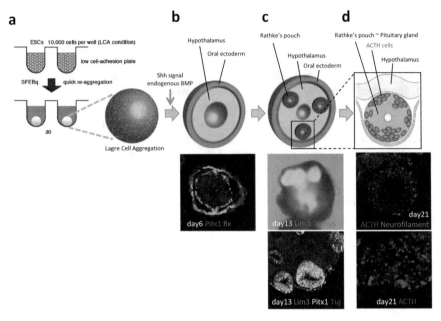

Fig. 2 In vitro differentiation into anterior pituitary from mouse ES cells (ESCs). (**a**) Diagram of SFEBq. (**b**) Two-layer formation in LCA aggregates. (**c**) Self-formation of Rathke's pouches. (**d**) Subsequent generation of ACTH⁺ cells

neuron differentiation instead of inducing oral ectodermal differentiation. The other condition was high-density cell aggregation (10,000 cells per aggregate instead of 3,000 in SFEBq culture), which we refer to as large cell aggregation (LCA; Fig. 2a). In the LCA culture, both the oral ectoderm (Pitx1/2+) and hypothalamic tissues co-existed within one aggregate (Fig. 2b).

LCA culture allows for the formation of oral ectoderm epithelium on the surface of mouse ES cell aggregates as well as hypothalamic neural tissue in the inner layer adjacent to the oral ectoderm (Fig. 2b). Treatment with a BMP4 antagonist, dorsomorphin, has been shown to suppress the generation of oral ectoderm (Suga et al. 2011). Quantitative polymerase chain reaction analyses revealed significantly higher internal BMP4 expression in LCA aggregates (Suga et al. 2011). Moreover, Koehler et al. succeeded in differentiating the otic placode (Fig. 1a; Koehler et al. 2013), which belongs to the head and oral ectoderm, following BMP treatment of mouse ES cells, which supports the reliability and robustness of this strategy. Our recent study showed that very low concentrations (picomolar level) of exogenous BMP4 treatment facilitated differentiation into non-neural ectoderms, which contained not only pituitary primordium but also dental germs (Ochiai et al. 2015). Taken together, these findings appear to indicate that appropriate BMP4 expression is important for head ectoderm induction (Wilson and Hemmati-Brivanlou 1995; Basch and Bronner-Fraser 2006; Davis and Camper 2007).

Self-Formation of Rathke's Pouch

In the developing embryo, Rathke's pouch forms at the midline of the head ectoderm. Shh is expressed in the ventral diencephalon and oral ectoderm but is excluded from the invaginating Rathke's pouch (Zhu et al. 2007; Wang et al. 2010). Rathke's pouch receives Shh signals from neighboring tissues in vivo, and Shh is known to provide positional information to adjust towards the midline (Zhu et al. 2007). Therefore, we added smoothened agonist (SAG) as a strong Shh signal to the differentiation medium of mouse ES cell aggregates in vitro. On day 13, multiple oval structures formed in the SAG-treated LCA SFEBq aggregates (Fig. 2c). The vesicles were situated between the oral ectoderm and hypothalamic neurons. Lim3 (formal gene name is Lhx3) expression indicated that the vesicles had similar characteristics to Rathke's pouch. These Lim3$^+$ tissues appeared as a thick epithelium on the surface that then invaginated and finally formed hollowed vesicles. The length of the major axis was about 200 μm, which is almost equal to the size of the embryonic Rathke's pouch. The size of Rathke's pouch seems to be prescribed.

Interactions between oral ectoderm and hypothalamic neurons appear to be critically important. Neither isolated surface ectoderm alone, nor isolated hypothalamic tissues alone, formed Lim3$^+$ pouches. Only in cases where the two divided components were re-assembled did Lim3$^+$ expression recover to some extent (Suga et al. 2011).

These findings demonstrate self-formation of Rathke's pouch in mouse ES cell aggregates. It has also been shown that Rathke's pouch forms even without mesenchymal cells, because this model contains only ectodermal cells.

Interestingly, a single aggregate often contains several pouches whereas there is usually only one pouch in the embryo (Suga et al. 2011). This finding suggests that several morphogenetic fields for pituitary placodes can be independently generated within the oral ectoderm epithelium on the surface of the ES cell aggregate, which is reminiscent of the *Vax1* knock-out mouse (Bharti et al. 2011). A second Rathke's pouch develops in addition to the orthotopic anlage in the *Vax1* knock-out mouse. Ectopic expression of FGF10, which is expressed in the infundibulum and implicated in pituitary induction, is also detected in the hypothalamic neuroepithelium overlying the second pouch. Thus, Vax1 likely limits the hypothalamic neuroepithelium area that generates pituitary-inducing signals. Indeed, *Vax1* expression in vivo is eliminated near the infundibulum, which has inducing activity for pituitary development. In the mouse ES aggregates used for pituitary differentiation in the present study, Vax1-positive cells did not exist in the hypothalamic area. Conversely, Wataya's aggregate for hypothalamic differentiation (Wataya et al. 2008) has been shown to contain Vax1-positive cells. We speculate that precise positioning in the hypothalamus slightly shifts as a result of BMP4 and Shh signals.

Differentiation Into Hormone-Producing Endocrine Cells

During pituitary development in the embryo, Lim3$^+$ pituitary progenitors commit to several lineages (Davis et al. 2011), i.e., corticotroph, somatotroph, lactotroph, thyrotroph, gonadotroph, and melanotroph lineages. Among them, the ACTH-producing corticotroph lineage expresses the transcription factor Tbx19 prior to ACTH expression. As Notch signaling inhibits Tbx19 expression (Lamolet et al. 2001; Zhu et al. 2006; Kita et al. 2007), we evaluated the effect of the Notch inhibitor DAPT. As a result, DAPT treatment increased Tbx19 expression in SAG-treated LCA SFEBq aggregates. A substantial number of ACTH$^+$ cells appeared in the Tbx19$^+$ lesion (Fig. 2d). Without DAPT treatment, corticotroph differentiation efficiency was decreased, and other lineages were not detected.

Previous reports have shown that canonical Wnt signaling promotes Pit1 expression (DiMattia et al. 1997; Olson et al. 2006; Sornson et al. 1996). Consistent with this finding, treatment with the Wnt agonist BIO increased Pit1 expression, resulting in subsequent GH$^+$ and PRL$^+$ cell differentiation.

Head mesenchyme has been suggested to promote pituitary development in vivo (Gleiberman et al. 1999). Therefore, we applied conditioned medium from PA6 stromal cells to SAG-treated LCA SFEBq aggregates. As a result, we successfully induced luteinizing hormone-positive, follicle-stimulating hormone-positive, and thyroid-stimulating hormone-positive cells. Further investigation is necessary to identify factors in the PA6-conditioned medium.

Lim3 is essential for these hormone-producing lineages. To suppress Lim3 expression in differentiating mouse ES cells, we used the Tet-inducible shRNA expression lentivirus vector system (kindly gifted from Hiroyuki Miyoshi at RIKEN BioResouce Center). Knockdown of Lim3 inhibited subsequent differentiation into hormone-producing cells, which supports altered pituitary development in Lim3 knockout mice (Sheng et al. 1996).

These results demonstrate the competence of ES cell-derived pituitary progenitors to generate multiple endocrine lineages in vitro.

Functionality of Induced ACTH$^+$ Cells

Positive and negative regulations by exogenous stimuli are characteristic for endocrine cells. To investigate in vitro functionality, we induced ACTH$^+$ cells for evaluation because they are most efficiently generated using the SAG-treated LCA SFEBq method.

After 10 min of stimulation by corticotropin releasing hormone (CRH), substantial amounts of ACTH were secreted from SAG-treated LCA SFEBq aggregates in vitro (Fig. 3a). The secreted ACTH concentration was similar to levels in mouse peripheral blood. ACTH secretion from the pituitary gland is negatively regulated

in vitro functionality

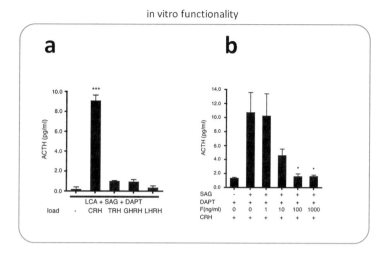

in vivo functionality

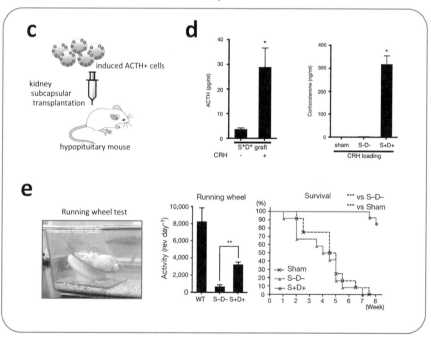

Fig. 3 Functional tests of mouse ES-derived ACTH$^+$ cells. (**a**) In vitro release from mouse ES-derived ACTH$^+$ cells. F, glucocorticoid pretreatment. Among the releasing factors, CRH most efficiently induces ACTH secretion. (**b**) Negative feedback test. Pretreatment with hydrocortisone suppresses CRH-stimulated ACTH secretion from aggregates. (**c**) In vivo functional test by ectopic transplantation. All mice, except for the WT mice, received a hypophysectomy; hypopituitarism was confirmed by CRH loading. S + D+, SAG- and DAPT-treated aggregates. S-D-, no SAG or DAPT treatment. The values shown on graphs represent mean ± s.e.m. *$P < 0.05$; **$P < 0.01$; ***$P < 0.001$ (modified from Suga et al. 2011). (**d**) Blood ACTH and subsequent release of corticosterone. (**e**) Improved activity and survival

by the downstream glucocorticoid hormone. Consistent with this control principle, in vitro ACTH secretion as a result of CRH stimulation was suppressed by glucocorticoid pre-treatment (Fig. 3b).

Similar to in vivo endocrine systems, these data demonstrate that mouse ES cell-derived ACTH$^+$ cells respond to both positive and negative regulators. These hormonal responses to surrounding regulators are indispensable for homeostasis. For this reason, the generation of anterior pituitary tissue that retains regulatory hormonal control in vitro is an important step for the development of cell transplantation therapies for pituitary diseases. Furthermore, we suggest that the endocrine organoid formed in this three-dimensional culture condition might better reflect the in vivo microenvironment. Such approaches may be beneficial for producing other functionally mature endocrine tissues.

Effect of Transplantation Into Hypophysectomized Model Animals

Finally, we evaluated the transplantation effect of the induced ACTH$^+$ cells. Because of technical difficulties, we chose ectopic transplantation into the kidney subcapsule (Fig. 3c) instead of orthotopic transplantation into the sella turcica. At one week after transplantation, blood ACTH levels were slightly, but significantly, increased. CRH loading induced a substantial elevation in blood ACTH levels (Fig. 3d). The downstream glucocorticoid hormone corticosterone was also significantly increased, indicating that ACTH from the graft sufficiently induced the downstream hormone (Fig. 3d).

Even without CRH loading, the basal levels of ACTH were higher. Importantly, corticosterone levels were also increased, suggesting that partial recovery of blood ACTH had a moderate, but biologically significant, effect (note that ED50 of the ACTH receptor MC2R for glucocorticoid production is around 9 pg/mL; Melmed 2011). In accordance with this finding, the treated hypophysectomized mice displayed higher spontaneous locomotor activities and survived significantly longer (Fig. 3e). Although CRH, which is secreted from the hypothalamus, should be diluted in the peripheral site, mouse ES cell derived pituitary tissues rescued survival and spontaneous activities, suggesting that basal secretion from these tissues was sufficient for those effects.

These findings showed that induced ACTH$^+$ cells derived from mouse ES cells acted as endocrine tissues and that regenerative medicine for pituitary dysfunction is feasible.

Adaptation to Human ES/iPS Cell Culture

The recovery of lost pituitary function is an important issue for medical studies because the anterior pituitary has poor potential for regeneration. Because some pituitary dysfunctions cannot be solely treated by drugs (Arima et al. 2014; Hahner et al. 2015; Sherlock et al. 2009), regenerative therapy employing stem cells should be considered as a new form of therapeutic intervention. Our SFEBq method (Suga et al. 2011) induces pituitary cells that can auto-regulate hormonal secretion and respond to changing circumstances. The application of this culture method to human ES cells is necessary for clinical purposes. However, poor survival of human ES cells in SFEB culture might limit the use of these cells for future medical applications. Our colleagues found that a selective Rho-associated kinase (ROCK) inhibitor, Y-27632, markedly diminished dissociation-induced apoptosis of human ES cells and enabled the cells to form aggregates in SFEB culture (Watanabe et al. 2007). Using this fundamentally important discovery, we attempted to adapt our pituitary-differentiating culture method for human ES cell culture. We were able to obtain corticotrophs and somatotrophs from the human ES cells, although these are still preliminary data.

Results demonstrated that the anterior pituitary self-forms in vitro following co-induction of the hypothalamic and oral ectoderm (Fig. 4a). The juxtaposition of these tissues facilitated the formation of the pituitary placode, and their features were consistent with characteristics of Rathke's pouch in vivo. The human ES cell-derived Rathke's pouch was much larger than the pouch formed by mouse ES cells, which was in accordance with the size difference between human and mouse embryonic Rathke's pouches. These pituitary placodes subsequently differentiated into pituitary hormone-producing cells. All six types of pituitary hormone-producing cells were identified (Fig. 4b). Among them, we confirmed that the human ES-derived corticotroph responded normally to releasing and feedback signals. Electron microscopy revealed secretory granules stored in the cytoplasm of these cells (Fig. 4c).

For both mouse and human ES cells, SFEB culture is a favorable method that can generate functional pituitary cells. Future studies will confirm whether human iPS cells can differentiate into pituitary cells using the same culture methods.

Future Perspectives

There are two primary uses for human ES/iPS cell-derived pituitary cells. One is the human model of development or disease. Results from our study showed that the present culture methods recapitulated embryogenesis, suggesting that it could be used in the area of developmental biology. In terms of diseases due to gene mutations, tissues derived from disease-specific iPS cells can be used for therapy screenings in a human disease model.

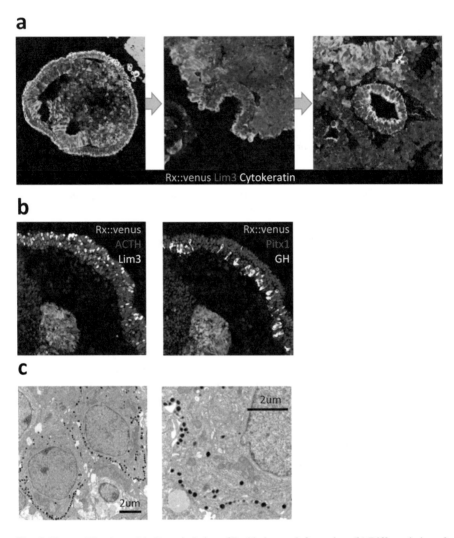

Fig. 4 Human ES culture. (**a**): Recapitulation of Rathke's pouch formation. (**b**) Differentiation of corticotrophs and somatotrophs. (**c**) Secretory granules characteristic of endocrine cells

The second major use for human ES/iPS cell-derived pituitary cells is for regenerative medicine. Although stem cell-based therapeutics provide high expectations for the treatment of diabetes mellitus, the use of regenerative medicine for hypothalamus-hypophyseal dysfunctions has received little attention.

ES cell-derived ACTH-producing cells function even after ectopic transplantation. This finding raises the possibility of relatively simple grafting of artificial ES/iPS cell-derived pituitary tissues into a peripheral site. These cells can function effectively if hormone secretion can be extrinsically controlled by releasing factors or small molecule agonists. However, ectopic transplantation is not perfect because physiological CRH released from the hypothalamus does not directly affect these grafts. Orthotopic transplantation of hormone-producing cells that are controlled by positive and negative regulators is one of the future candidates for complete therapy.

In future studies, it will be challenging to recapitulate an entire anterior pituitary gland that contains all endocrine components in three-dimensional cultures of human ES or iPS cells and to use such artificial pituitary tissues for orthotopic transplantation into the sella of a large mammal. To achieve this long-term goal, further studies are needed before pituitary regenerative medicine can be directly transferred to clinical use.

References

Andoniadou CL, Gaston-Massuet C, Reddy R, Schneider RP, Blasco MA, Le Tissier P, Jacques TS, Pevny LH, Dattani MT, Martinez-Barbera JP (2012) Identification of novel pathways involved in the pathogenesis of human adamantinomatous craniopharyngioma. Acta Neuropathol 124:259–271

Arima H, Wakabayashi T, Nagatani T, Fujii M, Hirakawa A, Murase T, Yambe Y, Yamada T, Yamakawa F, Yamamori I, Yamauchi M, Oiso Y (2014) Adipsia increases risk of death in patients with central diabetes insipidus. Endocr J 61:143–148

Babinet C, Cohen-Tannoudji M (2001) Genome engineering via homologous recombination in mouse embryonic stem (ES) cells: an amazingly versatile tool for the study of mammalian biology. An Acad Bras Cienc 73:365–383

Basch ML, Bronner-Fraser M (2006) Neural crest inducing signals. Adv Exp Med Biol 589:24–31

Ben-Shlomo A (2010) Pituitary gland: predictors of acromegaly-associated mortality. Nat Rev Endocrinol 6:67–69

Bernstein A, Breitman M (1989) Genetic ablation in transgenic mice. Mol Biol Med 6:523–530

Bharti K, Gasper M, Bertuzzi S, Arnheiter H (2011) Lack of the ventral anterior homeodomain transcription factor VAX1 leads to induction of a second pituitary. Development 138:873–878

Brinkmeier ML, Potok MA, Davis SW, Camper SA (2007) TCF4 deficiency expands ventral diencephalon signaling and increases induction of pituitary progenitors. Dev Biol 311:396–407

Chen J, Hersmus N, Van Duppen V, Caesens P, Denef C, Vankelecom H (2005) The adult pituitary contains a cell population displaying stem/progenitor cell and early embryonic characteristics. Endocrinology 146:3985–3998

Chen J, Gremeaux L, Fu Q, Liekens D, Van Laere S, Vankelecom H (2009) Pituitary progenitor cells tracked down by side population dissection. Stem Cells 27:1182–1195

Danjo T, Eiraku M, Muguruma K, Watanabe K, Kawada M, Yanagawa Y, Rubenstein JL, Sasai Y (2011) Subregional specification of embryonic stem cell-derived ventral telencephalic tissues by timed and combinatory treatment with extrinsic signals. J Neurosci 31:1919–1933

Davis SW, Camper SA (2007) Noggin regulates Bmp4 activity during pituitary induction. Dev Biol 305:145–160

Davis SW, Mortensen AH, Camper SA (2011) Birthdating studies reshape models for pituitary gland cell specification. Dev Biol 352:215–227

DiMattia GE, Rhodes SJ, Krones A, Carrière C, O'Connell S, Kalla K, Arias C, Sawchenko P, Rosenfeld MG (1997) The Pit-1 gene is regulated by distinct early and late pituitary-specific enhancers. Dev Biol 182:180–190

Eiraku M, Watanabe K, Matsuo-Takasaki M, Kawada M, Yonemura S, Matsumura M, Wataya T, Nishiyama A, Muguruma K, Sasai Y (2008) Self-organized formation of polarized cortical tissues from ESCs and its active manipulation by extrinsic signals. Cell Stem Cell 3:519–532

Eiraku M, Takata N, Ishibashi H, Kawada M, Sakakura E, Okuda S, Sekiguchi K, Adachi T, Sasai Y (2011) Self-organizing optic-cup morphogenesis in three-dimensional culture. Nature 472:51–56

Evans MJ, Kaufman MH (1981) Establishment in culture of pluripotential cells from mouse embryos. Nature 292:154–156

Fauquier T, Guérineau NC, McKinney RA, Bauer K, Mollard P (2001) Folliculostellate cell network: a route for long-distance communication in the anterior pituitary. Proc Natl Acad Sci U S A 98:8891–8896

Fu Q, Gremeaux L, Luque RM, Liekens D, Chen J, Buch T, Waisman A, Kineman R, Vankelecom H (2012) The adult pituitary shows stem/progenitor cell activation in response to injury and is capable of regeneration. Endocrinology 153:3224–3235

Garcia-Lavandeira M, Saez C, Diaz-Rodriguez E, Perez-Romero S, Senra A, Dieguez C, Japon MA, Alvarez CV (2012) Craniopharyngiomas express embryonic stem cell markers (SOX2, OCT4, KLF4, and SOX9) as pituitary stem cells but do not coexpress RET/GFRA3 receptors. J Clin Endocrinol Metab 97:E80–87

Gaston-Massuet C, Andoniadou CL, Signore M, Jayakody SA, Charolidi N, Kyeyune R, Vernay B, Jacques TS, Taketo MM, Le Tissier P, Dattani MT, Martinez-Barbera JP (2011) Increased Wingless (Wnt) signaling in pituitary progenitor/stem cells gives rise to pituitary tumors in mice and humans. Proc Natl Acad Sci U S A 108:11482–11487

Gleiberman AS, Fedtsova NG, Rosenfeld MG (1999) Tissue interactions in the induction of anterior pituitary: role of the ventral diencephalon, mesenchyme, and notochord. Dev Biol 213:340–353

Gremeaux L, Fu Q, Chen J, Vankelecom H (2012) Activated phenotype of the pituitary stem/progenitor cell compartment during the early-postnatal maturation phase of the gland. Stem Cells Dev 21:801–813

Hahner S, Spinnler C, Fassnacht M, Burger-Stritt S, Lang K, Milovanovic D, Beuschlein F, Willenberg HS, Quinkler M, Allolio B (2015) High incidence of adrenal crisis in educated patients with chronic adrenal insufficiency: a prospective study. J Clin Endocrinol Metab 100:407–416

Ikeda H, Osakada F, Watanabe K, Mizuseki K, Haraguchi T, Miyoshi H, Kamiya D, Honda Y, Sasai N, Yoshimura N, Takahashi M, Sasai Y (2005) Generation of Rx+/Pax6+ neural retinal precursors from embryonic stem cells. Proc Natl Acad Sci U S A 102:11331–11336

Kadoshima T, Sakaguchi H, Nakano T, Soen M, Ando S, Eiraku M, Sasai Y (2013) Self-organization of axial polarity, inside-out layer pattern, and species-specific progenitor dynamics in human ES cell-derived neocortex. Proc Natl Acad Sci U S A 110:20284–20289

Kawasaki H, Suemori H, Mizuseki K, Watanabe K, Urano F, Ichinose H, Haruta M, Takahashi M, Yoshikawa K, Nishikawa S, Nakatsuji N, Sasai Y (2002) Generation of dopaminergic neurons and pigmented epithelia from primate ES cells by stromal cell-derived inducing activity. Proc Natl Acad Sci U S A 99:1580–1585

Kikuchi M, Yatabe M, Kouki T, Fujiwara K, Takigami S, Sakamoto A, Yashiro T (2007) Changes in E- and N-cadherin expression in developing rat adenohypophysis. Anat Rec 290:486–490

Kita A, Imayoshi I, Hojo M, Kitagawa M, Kokubu H, Ohsawa R, Ohtsuka T, Kageyama R, Hashimoto N (2007) Hes1 and Hes5 control the progenitor pool, intermediate lobe specification, and posterior lobe formation in the pituitary development. Mol Endocrinol 21:1458–1466

Koehler KR, Mikosz AM, Molosh AI, Patel D, Hashino E (2013) Generation of inner ear sensory epithelia from pluripotent stem cells in 3D culture. Nature 500:217–221

Lamolet B, Pulichino AM, Lamonerie T, Gauthier Y, Brue T, Enjalbert A, Drouin J (2001) A pituitary cell-restricted T box factor, Tpit, activates POMC transcription in cooperation with Pitx homeoproteins. Cell 104:849–859

Langlais D, Couture C, Kmita M, Drouin J (2013) Adult pituitary cell maintenance: lineage-specific contribution of self-duplication. Mol Endocrinol 27:1103–1112

Li H, Collado M, Villasante A, Matheu A, Lynch CJ, Cañamero M, Rizzoti K, Carneiro C, Martínez G, Vidal A, Lovell-Badge R, Serrano M (2012) p27(Kip1) directly represses Sox2 during embryonic stem cell differentiation. Cell Stem Cell 11:845–852

Luque RM, Lin Q, Córdoba-Chacón J, Subbaiah PV, Buch T, Waisman A, Vankelecom H, Kineman RD (2011) Metabolic impact of adult-onset, isolated, growth hormone deficiency (AOiGHD) due to destruction of pituitary somatotropes. PLoS One 6, e15767

Melmed S (ed) (2011) The pituitary, 3rd edn. Academic Press, Cambridge, MA, p 61

Mollard P, Hodson DJ, Lafont C, Rizzoti K, Drouin J (2012) A tridimensional view of pituitary development and function. Trends Endocrinol Metab 23:261–269

Morizane A, Takahashi J, Shinoyama M, Ideguchi M, Takagi Y, Fukuda H, Koyanagi M, Sasai Y, Hashimoto N (2006) Generation of graftable dopaminergic neuron progenitors from mouse ES cells by a combination of coculture and neurosphere methods. J Neurosci Res 83:1015–1027

Muguruma K, Nishiyama A, Ono Y, Miyawaki H, Mizuhara E, Hori S, Kakizuka A, Obata K, Yanagawa Y, Hirano T, Sasai Y (2010) Ontogeny-recapitulating generation and tissue integration of ES cell-derived Purkinje cells. Nat Neurosci 13:1171–1180

Nakano T, Ando S, Takata N, Kawada M, Muguruma K, Sekiguchi K, Saito K, Yonemura S, Eiraku M, Sasai Y (2012) Self-formation of optic cups and storable stratified neural retina from human ESCs. Cell Stem Cell 10:771–785

Ochiai H, Suga H, Yamada T, Sakakibara M, Kasai T, Ozone C, Ogawa K, Goto M, Banno R, Tsunekawa S, Sugimura Y, Arima H, Oiso Y (2015) BMP4 and FGF strongly induce differentiation of mouse ES cells into oral ectoderm. Stem Cell Res 15:290–298

Olson LE, Tollkuhn J, Scafoglio C, Krones A, Zhang J, Ohgi KA, Wu W, Taketo MM, Kemler R, Grosschedl R, Rose D, Li X, Rosenfeld MG (2006) Homeodomain-mediated beta-catenin-dependent switching events dictate cell-lineage determination. Cell 125:593–605

Osakada F, Ikeda H, Mandai M, Wataya T, Watanabe K, Yoshimura N, Akaike A, Sasai Y, Takahashi M (2008) Toward the generation of rod and cone photoreceptors from mouse, monkey and human embryonic stem cells. Nat Biotechnol 26:215–224

Sakaguchi H, Kadoshima T, Soen M, Narii N, Ishida Y, Ohgushi M, Takahashi J, Eiraku M, Sasai Y (2015) Generation of functional hippocampal neurons from self-organizing human embryonic stem cell-derived dorsomedial telencephalic tissue. Nat Commun 6:8896

Sasai Y, Eiraku M, Suga H (2012) In vitro organogenesis in three dimensions: self-organising stem cells. Development 139:4111–4121

Sheng HZ, Zhadanov AB, Mosinger B Jr, Fujii T, Bertuzzi S, Grinberg A, Lee EJ, Huang SP, Mahon KA, Westphal H (1996) Specification of pituitary cell lineages by the LIM homeobox gene Lhx3. Science 272:1004–1007

Sherlock M, Reulen RC, Alonso AA, Ayuk J, Clayton RN, Sheppard MC, Hawkins MM, Bates AS, Stewart PM (2009) ACTH deficiency, higher doses of hydrocortisone replacement, and radiotherapy are independent predictors of mortality in patients with acromegaly. J Clin Endocrinol Metab 94:4216–4223

Sornson MW, Wu W, Dasen JS, Flynn SE, Norman DJ, O'Connell SM, Gukovsky I, Carrière C, Ryan AK, Miller AP, Zuo L, Gleiberman AS, Andersen B, Beamer WG, Rosenfeld MG (1996)

Pituitary lineage determination by the Prophet of Pit-1 homeodomain factor defective in Ames dwarfism. Nature 384:327–333

Suga H, Kadoshima T, Minaguchi M, Ohgushi M, Soen M, Nakano T, Takata N, Wataya T, Muguruma K, Miyoshi H, Yonemura S, Oiso Y, Sasai Y (2011) Self-formation of functional adenohypophysis in three-dimensional culture. Nature 480:57–62

Takuma N, Sheng HZ, Furuta Y, Ward JM, Sharma K, Hogan BL, Pfaff SL, Westphal H, Kimura S, Mahon KA (1998) Formation of Rathke's pouch requires dual induction from the diencephalon. Development 125:4835–4840

Wang Y, Martin JF, Bai CB (2010) Direct and indirect requirements of Shh/Gli signaling in early pituitary development. Dev Biol 348:199–209

Watanabe K, Kamiya D, Nishiyama A, Katayama T, Nozaki S, Kawasaki H, Watanabe Y, Mizuseki K, Sasai Y (2005) Directed differentiation of telencephalic precursors from embryonic stem cells. Nat Neurosci 8:288–296

Watanabe K, Ueno M, Kamiya D, Nishiyama A, Matsumura M, Wataya T, Takahashi JB, Nishikawa S, Nishikawa S, Muguruma K, Sasai Y (2007) A ROCK inhibitor permits survival of dissociated human embryonic stem cells. Nat Biotechnol 25:681–686

Wataya T, Ando S, Muguruma K, Ikeda H, Watanabe K, Eiraku M, Kawada M, Takahashi J, Hashimoto N, Sasai Y (2008) Minimization of exogenous signals in ES cell culture induces rostral hypothalamic differentiation. Proc Natl Acad Sci USA 105:11796–11801

Wilson PA, Hemmati-Brivanlou A (1995) Induction of epidermis and inhibition of neural fate by Bmp-4. Nature 376:331–333

Zhu X, Zhang J, Tollkuhn J, Ohsawa R, Bresnick EH, Guillemot F, Kageyama R, Rosenfeld MG (2006) Sustained Notch signaling in progenitors is required for sequential emergence of distinct cell lineages during organogenesis. Genes Dev 20:2739–2753

Zhu X, Gleiberman AS, Rosenfeld MG (2007) Molecular physiology of pituitary development: signaling and transcriptional networks. Physiol Rev 87:933–963

Regulation of Body Weight and Metabolism by Tanycyte-Derived Neurogenesis in Young Adult Mice

Seth Blackshaw, Daniel A. Lee, Thomas Pak, and Sooyeon Yoo

Abstract The hypothalamus controls many homeostatic and instinctive physiological processes, including the sleep-wake cycle, food intake, and sexually dimorphic behaviors. These behaviors are regulated by environmental and physiological cues, although the molecular and cellular mechanisms that underlie these effects are still poorly understood. Recently, it has become clear that both the juvenile and adult hypothalamus exhibit neurogenesis, which modifies homeostatic neural circuitry. In this manuscript, we report data addressing the role of sex-specific and dietary factors in controlling neurogenesis in the mediobasal hypothalamus. We report that a high fat diet (HFD) activates neurogenesis in the median eminence (ME) of young adult female, but not male mice, and that focal irradiation of the ME in HFD-fed mice reduces weight gain in females, but not males. These results suggest that some physiological effects of HFD are mediated by sexually dimorphic neurogenesis in the ME. We present these findings in the context of other studies on

S. Blackshaw (✉)
Solomon H. Snyder Department of Neuroscience, Johns Hopkins University School of Medicine, Baltimore, MD, USA

Institute for Cell Engineering, Johns Hopkins University School of Medicine, Baltimore, MD, USA

Department of Ophthalmology, Johns Hopkins University School of Medicine, Baltimore, MD, USA

Department of Neurology, Johns Hopkins University School of Medicine, Baltimore, MD, USA

Center for Human Systems Biology, Johns Hopkins University School of Medicine, Baltimore, MD, USA
e-mail: sblack@jhmi.edu

D.A. Lee
Solomon H. Snyder Department of Neuroscience, Johns Hopkins University School of Medicine, Baltimore, MD, USA

Division of Biology and Biomedical Engineering, California Institute of Technology, Pasadena, CA, USA

T. Pak • S. Yoo
Solomon H. Snyder Department of Neuroscience, Johns Hopkins University School of Medicine, Baltimore, MD, USA

© The Author(s) 2016
D. Pfaff, Y. Christen (eds.), *Stem Cells in Neuroendocrinology*, Research and Perspectives in Endocrine Interactions, DOI 10.1007/978-3-319-41603-8_5

51

the cellular and molecular mechanisms that regulate neurogenesis in postnatal and adult hypothalamus.

Introduction

Obesity and metabolic disorders are severe public health problems in developed countries. The pathophysiological effects of metabolic disease are partially mediated by hypothalamic inflammation (Thaler et al. 2012; Cai 2013; Purkayastha and Cai 2013) and by compensatory changes in hypothalamic neural circuitry triggered by obesity-induced neural injury. Recent studies have revealed hypothalamic neural injury in obese patients (Thaler et al. 2012). An understanding of the cellular and molecular responses to hypothalamic injury induced by dietary factors may identify new therapeutic targets for treating obesity and metabolic disorders (Lee and Blackshaw 2012).

Newborn neurons in the postnatal and adult hypothalamus have been described in multiple vertebrate species, including zebrafish (Wang et al. 2012; McPherson et al. 2016), chick (Kisliouk et al. 2014), hamster (Mohr and Sisk 2013), mouse (Lee et al. 2012), rats (Matsuzaki et al. 2015), and sheep (Batailler et al. 2015). This finding suggests a degree of plasticity that is evolutionarily conserved and likely extends to humans as well (Batailler et al. 2013; Dahiya et al. 2011; Noguiera et al. 2014). Both juvenile and adult mammalian hypothalamus exhibit ongoing neurogenesis that can be regulated by growth and differentiation factors (Pencea et al. 2001; Xu et al. 2005; Kokoeva et al. 2005; Perez-Martin et al. 2010; Robins et al. 2013b), diet (Lee et al. 2012; Li et al. 2012; McNay et al. 2012; Gouaze et al. 2013; Bless et al. 2014), antidepressants (Sachs and Caron 2014), exercise (Matsuzaki et al. 2015), and hormones (Ahmed et al. 2008; Bless et al. 2014). Although these studies generally agree that levels of constitutive hypothalamic neurogenesis are much lower than those seen in well-characterized neurogenic zones in the adult brain, such as the subventricular zone of the lateral ventricles or the subgranular zone of the lateral hypothalamus (Lee et al. 2012), they often report differing effects of extrinsic factors on cell proliferation and neurogenesis in hypothalamus. In addition, these studies make opposing claims about levels of neurogenesis and proliferation in certain hypothalamic regions, and the cell(s) of origin for these adult-born neurons remain controversial (Lee and Blackshaw 2012).

For instance, using a combination of in vitro cell culture and in vivo genetic lineage analysis, studies have claimed that a population of Sox2-positive (Li et al. 2012) and/or NG2-positive progenitors in the mediobasal hypothalamic parenchyma (Robins et al. 2013b) acts as multipotent neural progenitors. Tanycytes of the hypothalamic ventricular zone have also been reported to act as neural progenitors (Xu et al. 2005; Lee et al. 2012; Li et al. 2012; Haan et al. 2013; Robins et al. 2013a), and it has been variously claimed that dorsal located alpha2 and ventral beta2 tanycytes of the median eminence show the greatest levels of neurogenic potential (Robins et al. 2013a; Lee et al. 2012).

Other studies have reported that a range of extrinsic factors, such as dietary and hormonal signals as well as growth and differentiation factors, can also modulate postnatal hypothalamic neurogenesis. High-fat diet (HFD) has been reported to constitutively inhibit neurogenesis in the mediobasal hypothalamic parenchyma (Li et al. 2012; McNay et al. 2012) while activating neurogenesis in median eminence (ME; Lee et al. 2012; Hourai and Miyata 2013). It has also been reported that neurogenesis occurs in a sexually dimorphic pattern during puberty in hypothalamic regions, such as the preoptic area and anterioventral paraventricular nucleus that control sexual behavior (Ahmed et al. 2008), although the source of these young adult-generated neurons was not investigated.

Although these results seem discrepant at first glance, a closer examination reveals that these observed effects may result from methodological differences among the studies (Lee et al. 2012; Migaud et al. 2010). For instance, while multiple groups have reported that long-term administration of HFD inhibits cell proliferation and neurogenesis in hypothalamic parenchyma (Li et al. 2012; McNay et al. 2012; Gouaze et al. 2013; Bless et al. 2014), studies investigating acute responses to HFD have reported increased hypothalamic cell proliferation and neurogenesis (Thaler et al. 2012; Gouaze et al. 2013). Acute HFD administration has also been reported to rapidly induce hypothalamic inflammation, increasing cytokine signaling (Thaler et al. 2012). The physiological response in acute versus chronic HFD administration may serve different, but equally important, roles in maintaining metabolic homeostasis.

Hypothalamic progenitor cell populations may likewise respond differentially, and in some cases with opposite reactions, to dietary signals such as HFD. The ME, for instance, lies outside the blood–brain barrier; it is thus exposed to higher effective concentrations of circulating dietary and hormonal cues than the hypothalamic parenchyma (Fry et al. 2007; Langlet et al. 2013b). In contrast, all tanycyte subtypes directly contact the cerebrospinal fluid (CSF) and can potentially respond to intracerebral ventricular signals (Bennett et al. 2009; Bolborea and Dale 2013).

The age of the mice used for these studies has ranged from early postnatal (Lee et al. 2012) to young adult (Lee et al. 2012; Ahmed et al. 2008), to 3–12 months of age (Lee et al. 2012; Kokoeva et al. 2005; McNay et al. 2012; Kokoeva et al. 2007). Finally, studies of postnatal and adult neurogenesis in the ventrobasal hypothalamus have typically used either only male (Kokoeva et al. 2005, 2007; McNay et al. 2012; Li et al. 2012; Sachs and Caron 2014) or only female (Lee et al. 2012; Bless et al. 2014) mice. Neurogenesis in other hypothalamic regions is sexually dimorphic (Ahmed et al. 2008), making this but one additional methodological difference that could contribute to differences in the levels, location and dietary regulation of hypothalamic neurogenesis reported in these studies.

To clarify the extent to which sex-dependent factors might regulate neurogenesis in different hypothalamic regions, we investigated levels of hypothalamic neurogenesis in both the arcuate nucleus (ArcN) and ME in male and female young adult mice fed normal chow and HFD. We also investigated the effects of low-protein diet (LPD) and caloric restriction (CR) in these same areas in female mice. These dietary treatments led to significant and region-specific differences in neurogenesis. Most notably, HFD treatment inhibited ArcN neurogenesis in both sexes while selectively stimulating ME neurogenesis in female mice. In mice fed HFD, we found that inhibiting ME neurogenesis by computer tomography-guided focal irradiation attenuated weight gain in females but not males. These findings advance our understanding of extrinsic factors that regulate adult hypothalamic neurogenesis and reconcile a number of seemingly discrepant recent studies on this topic.

Material and Methods

Animals

Five week old female or male C57BL/6 mice were obtained from Charles River and housed in a 14/10-h light–dark cycle with free access to normal chow (Teklad F6 Rodent Diet 8664:: Protein (kcal): 31 %, Carbohydrate (kcal): 50 %, Fat (kcal): 19 %, Harlan Teklad, Madison, WI) and water. Where indicated, animals were provided with a high-fat diet (HFD) (Catalog #: D12492i:: Protein (kcal): 20 %, Carbohydrate (kcal): 20 %, Fat (kcal): 60 %, Research Diets, New Brunswick, NJ) or low protein diet (Catalog#: D11112203:: Protein (kcal): 8 %, Carbohydrate (kcal): 76 %, Fat (kcal): 16 %, Research Diets, New Brunswick, NJ). All mice used in these studies were maintained and euthanized according to protocols approved by the Institutional Animal Care and Use Committee at the Johns Hopkins School of Medicine.

Caloric Restriction

Five-week-old female C57BL/6 mice were obtained from Charles River and put on a high-fat diet (HFD: 60 % kcals from fat, Research Diets #D12492). At six weeks old, mice were separated into two groups: (control group) HFD ad libitum and HFD caloric restriction (CR). CR is at 70 % of the HFD control group's average food intake. This was calculated by providing the amount of the control group's average food intake, plus an amount equal to the standard error of that group's intake, to ensure that mice would have enough food both to eat, and to spill, and maintain 70 % of the HFD control group's average food intake. Food intakes were measured twice per week and used to calculate the CR levels to be used for 0.5 weeks until the next food intake assessment.

Reagents

Bromodeoxyuridine (BrdU)

Where indicated, young adult mice received bromodeoxyuridine (BrdU; Sigma) administrated in the morning and evening by intraperitoneal injection at 50 mg/kg of body weight from P45 to P53.

Tissue Processing and Antibodies

Adult mice were sacrificed, perfused with 4 % PFA/PBS, and cryoprotected as previously described (Lee *et al.* 2012). Serial sections (40 μm thick) were collected and stored at −20 ° C. Free-floating sections were immunostained using the following primary antibodies and working concentrations: mouse monoclonal anti-phospho-H2AX, Ser139, clone JBW301 (1:700, Millipore), rat monoclonal anti-BrdU (1:200, Accurate, Westbury, NY), mouse monoclonal anti-Hu (5 μg/ml, Molecular Probes, Carlsbad, CA). Double staining was visualized with Alexa Fluor 555-, and Alexa Fluor 488 (1:500, Molecular Probe, Carlsbad, CA). 4′,6-diamidino-2-phenylindole (DAPI) was used as a nuclear counterstain.

Immunohistochemistry

γH2AX immunostaining was performed as previously described (Lee et al. 2013). For BrdU immunostaining, sections were first incubated in 2 N HCl at 37 °C for 30 min, and rinsed in 0.1 M boric acid (pH 8.5) at room temperature for 10 min. Sections were then rinsed in PBST, blocked for 5 min in SuperBlock (ScyTek), and incubated overnight with in anti-BrdU antibody in 5 % normal horse serum in PBS/0.16 % Triton X-100 at 4 °C in blocking solution. Sections were washed in PBST, incubated with secondary antibodies in blocking solution at RT for 2 h, washed in PBST, mounted on Superfrost Plus slides (Fisher, Hampton, NH), and coverslipped with Gelvatol mounting medium.

Cell Quantification

All tissue sections used for quantification were imaged using confocal microscopy (Meta 510, Zeiss Microscopy). ME cells were counted. The dorsal-ventral boundary of the cells counted was the third ventricle (3 V) floor and the ventral edge of the external layer of the ME. The lateral boundaries were a 20 μm medial inset off the corner of the 3 V. ME dorsal and ventral boundaries remained identical to as previously described. Seven 40-μm coronal serial sections (280 μm) were counted between -1.515 mm and -1.875 mm from Bregma. On the rare occasion, a section would be lost in the collection process. If available, the next section in the mouse sample was taken and counted (seven sections were counted total). For analysis of newborn Hu^+ neurons, for each section analyzed, Hu^+DAPI^+ and $Hu^+BrdU^+DAPI^+$ neurons within the ME were counted in the region defined above, excluding cells of the uppermost focal plane to avoid oversampling. To determine the frequency of $BrdU^+$ cells expressing Hu, dual fluorescence-labeled sections were examined by confocal microscopy using a $20\times$ objective and $1.5\times$ digital zoom. For each marker and treatment condition, seven representative serial sections from each animal were examined. Sections were scored for double labeling by manual examination of optical slices. Cells were considered positive for a given phenotypic marker when the marker-specific labeling was unambiguously associated with a $BrdU^+$ nucleus. Cells were spot-checked in all three dimensions by Z-stack using a 63x objective. Images of $Hu^+BrdU^+DAPI^+$ labeling in feeding conditions (Fig. 1b) were blinded prior to counting. Cell counts are described in the text and figure legends as mean of several samples $\pm$ s.e.m., total cell counts, and the number of samples examined to derive those total cell counts.

Focal Irradiation of Ventrobasal Hypothalamus

Radiation (10 Gy) was delivered using the Small Animal Radiation Research Platform (SARRP), a dedicated laboratory focal radiation device with CT capabilities (Xstrahl, Inc.). A detailed video and protocol describing this focal irradiation methodology is available online (http://www.jove.com/video/50716/functional-interrogation-adult-hypothalamic-neurogenesis-with-focal) (Lee et al. 2013). Sham controls were performed in parallel. Mice in this cohort were caged, transported to the procedure room, received the same anesthesia, and received similar amounts of ambient radiation coming from the SARRP as the irradiated cohort. Sham controls differed only in that they did not receive a direct radiation beam and CT scan.

Longitudinal Collection of Weight Data

Weight data from each mouse subject was collected at the time of sham or irradiation treatment. Longitudinal weight gain was normalized to weight at the time of treatment for each subject. Weight data was taken every 0.5 weeks for each mouse subject. Female mice were group housed. Male mice were all group-housed initially, but were separated if they were observed to fight. To reduce variation resulting from changes in housing, the numbers of male mice that were individually and group housed were matched at all times between the sham and irradiated cohorts. For one experiment, HFD-fed female sham or irradiated cohorts, blood

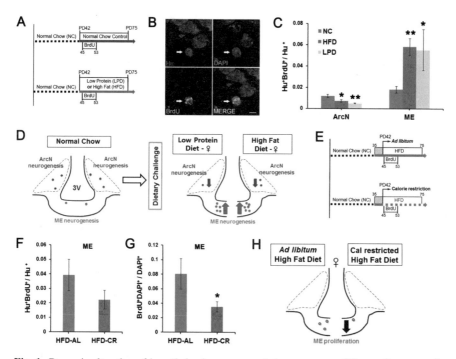

Fig. 1 Dynamic alteration of hypothalamic neurogenesis in response to different dietary conditions. (**a**) Experimental design schematic. Female mice were either continuously fed on normal chow (NC) or switched to the low protein diet (LPD) or high fat diet (HFD) at postnatal day (PD) 42. After 3 days, BrdU was injected intraperitoneally twice per day for 9 days. Mice were sacrificed 1 month after the onset of BrdU administration. (**b**) Representative high magnification image of BrdU and Hu double-positive P75 adult-born neurons (*white arrows*) located in median eminence (ME) of female mice fed HFD. (**c**) Quantitative comparison of diet-dependent neurogenesis (Hu$^+$BrdU$^+$/Hu$^+$ neurons) in the arcuate nucleus (ArcN) and ME. (**d**) Schematic summarizing opposite effects of dietary change on neurogenesis between ArcN and ME. (**e**) Scheme of experimental design for calorie restriction. After 1 week initial adjustment to HFD (*red striped square*), female mice were fed either ad libitumHFD or were calorie restricted (CR) on the HFD to 70 % of the ad libitum-fed mice, from PD42 onward. After 3 days, BrdU was injected intraperitoneally twice per day for 9 days. Mice were sacrificed 1 month after BrdU administration onset. (**f, g**) Quantitative comparison of ME neurogenesis (Hu$^+$BrdU$^+$/Hu$^+$ neurons) or proliferation (BrdU$^+$DAPI$^+$/DAPI$^+$ cells) in CR HFD-fed (HFD-CR) mice or ad libitum HFD-fed (HFD-AL) mice. (**h**) Schematic summarizing significant reduction of cell proliferation in the ME of mice fed on HFD-CR. *P < 0.05, **P < 0.003. Scale bar: 5 μm

samples were collected 1 week following treatment and a standard complete blood count panel was taken. There was no statistically significant difference in any of blood components.

Statistical Analysis

Figures are shown as mean ± standard error of the mean. Two-tailed Student's *t*-test was applied. A p-value ≤ 0.05 indicated significant group difference.

Results

Dietary Signals Differentially Regulate Neurogenesis and Cell Proliferation in ME and ArcN

Our group previously demonstrated that feeding young adult female mice HFD led to significantly increased ME neurogenesis (Lee et al. 2012). We set out to test whether additional dietary conditions could also alter neurogenesis in the ME and whether these conditions led to comparable changes in neurogenesis in the ArcN, which lies inside the blood–brain barrier within the hypothalamus proper. Young adult female mice were switched onto control normal chow (NC), HFD or LPD beginning at postnatal day (P) 42. Cell proliferation was tracked using twice-daily intraperitoneal (i.p.) injections of BrdU from P45-53 (Fig. 1a). Mice were euthanized at P75, and brains were immunostained for BrdU and the pan-neuronal marker HuC/D. The fraction of Hu$^+$ cells that were also BrdU$^+$ was quantified to assess levels of neurogenesis.

Baseline levels of hypothalamic neurogenesis [(Hu$^+$BrdU$^+$)/Hu$^+$ neurons] in mice fed NC were low and did not differ significantly between the two regions (NC ArcN [0.012 ± 0.002, n = 5] vs NC ME [0.015 ± 0.004, n = 7], p = 0.47, equal variance; Fig. 1c). Both HFD-fed mice (0.0072 ± 0.0012, n = 5, p = 0.044, equal "n") and LPD-fed mice (0.0048 ± 0.0004, n = 5, p = 0.0022, equal "n") showed a substantial reduction in the fraction of Hu$^+$BrdU$^+$ ArcN neurons. In contrast, both HFD-fed mice (0.058 ± 0.008, n = 9, p = 0.0005, unequal "n," equal variance) and LPD-fed mice (0.055 ± 0.019, n = 4, p = 0.025, unequal "n," equal variance) showed a significant increase in the fraction of Hu$^+$BrdU$^+$ ME neurons (Fig. 1c). The differences in neurogenesis levels between the ArcN and ME following both HFD (ArcN [0.0072 ± 0.001, n = 5] vs. ME [0.058 ± 0.008, n = 9], p = 0.0005, equal variance) and LPD (ArcN [0.0048 ± 0.0004, n = 5] vs. ME [0.055 ± .019, n = 4], p = 0.021, equal variance] were significant and implied that neural progenitor populations in these two regions responded differentially to these dietary cues (Fig. 1c, d). The HFD-induced inhibition of ArcN neurogenesis was similar to observations of adult male mice made by other groups (McNay et al. 2012; Li et al. 2012).

Because previous studies reported that CR could reverse the effects of HFD on ArcN neurogenesis (McNay et al. 2012), we next tested whether CR could likewise modulate HFD-induced ME neurogenesis. For these studies, female mice were housed individually and allowed either ad libitum or restricted access to HFD

starting at P45. Restricted HFD access was at 70 % of the caloric intake of animals fed ad libitum (Fig. 1e). BrdU labeling and immunohistochemistry were conducted as described above. CR HFD-fed mice trended towards a decrease in neurogenesis levels in ME (HFD ad lib [0.039 ± 0.011, n = 4] vs HFD-CR [0.022 ± 0.006, n = 6] p = 0.19, equal variance), but this effect did not reach significance (Fig. 1f). However, we observed that overall BrdU incorporation in ME cells was significantly reduced (HFD ad lib [0.08 ± 0.02, n = 4] vs HFD-CR [0.035 ± 0.007, n = 6], p = .043, equal variance) (Fig. 1g). We thus conclude that CR inhibits proliferation of progenitor cells within the ME (Fig. 1h).

Sex-Specific Differences in Diet-Induced Hypothalamic Neurogenesis

Since these studies were all performed in young adult females, we next tested whether the levels of baseline and HFD-induced ME neurogenesis were different between the sexes. We used diet and BrdU labeling conditions identical to those in Fig. 1a to measure levels of neurogenesis in male mice fed either NC or HFD. We observed that young adult males fed NC showed levels of neurogenesis in both ArcN and ME that were low and not significantly different from age-matched females (NC-fed: Male ArcN [0.0119 ± 0.0012, n = 3] vs Male ME [0.020 ± 0.009, n = 3], p = 0.40, equal variance; Fig. 2a). We likewise observed that male mice fed HFD showed reduced levels of ArcN neurogenesis relative to male mice fed NC (NC ArcN [0.0119 ± 0.0016, n = 3] vs HFD ArcN [0.0070 ± 0.0011, n = 5] p = 0.036, equal variance). However, in sharp contrast to what we observed in females, male mice showed no increase in ME neurogenesis when fed HFD (0.020 ± 0.009, n = 3, vs 0.025 ± 0.006, n = 4: NC vs HFD, p = 0.36; Fig. 2a), implying that neural progenitor populations in the ME showed sex-specific increases in neurogenesis in response to HFD (Fig. 2b).

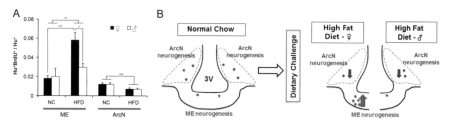

Fig. 2 Sex differences in neurogenic response to HFD. (**a**) Quantification of hypothalamic neurogenesis (Hu$^+$BrdU$^+$/Hu$^+$ neurons) in the ArcN and ME of young adult male mice fed HFD or NC. (**b**) Schematic summarizing sexual dimorphism of dietary challenge on ME neurogenesis

Blocking Neurogenesis in ME Attenuates HFD-Induced Weight Gain in Young Adult Female, But Not Male, Mice

We previously demonstrated that computer tomography-guided focal irradiation could selectively inhibit cell proliferation in the ME while sparing proliferation in ArcN (Lee et al. 2012, 2013). Our radiological approach reduced ME neurogenesis by ~85% (Lee et al. 2012), in line with previous approaches that used focal irradiation in other mammalian neurogenic niches (Ford et al. 2011). We next tested whether selective radiological inhibition of ME neurogenesis (Fig. 3a) in both males and female mice fed HFD led to sex-specific differences in regulation of body weight. The specificity of the focal irradiation was demonstrated using γH2AX immunostaining (Fig. 3b; Lee et al. 2012, 2013). Longitudinal body weight measurements were then taken for male and female mice that underwent either focal irradiation or sham treatment and that were fed either NC or HFD. Weight changes were normalized to the starting weight of each mouse at the time of sham or irradiation treatment.

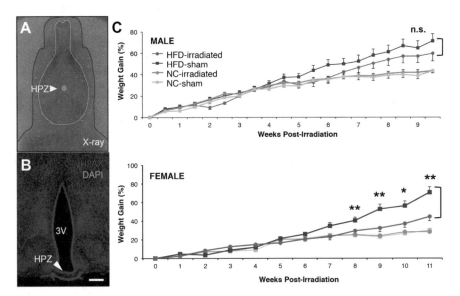

Fig. 3 Sex differences in diet-induced weight gain after focal inhibition of ME neurogenesis. Mice received either NC or HFD at 5 weeks of age. Sham treatment or 10 Gy of computer tomography-guided focal radiation was applied to the ME of young adult mice (P42) as previously described (Lee et al. 2013). (**a**) Superimposition of dosimetry-film acquired with 1-mm radiation beam in phantom with an X-ray of a real mouse subject (*blue line*). *White circle* (*arrow*) indicates 10-Gy radiation dose focally targeted to hypothalamic proliferative zone (HPZ). (**b**) Confirmation of radiation targeting accuracy by γH2AX immunostaining, an indicator of DNA double strand breaks and radiation localization. (**c**) Weight gain was assessed from the time of either sham or focal irradiation treatment in NC- and HFD-fed mice of both sexes

We observed no long-term changes in body weight between sham and irradiated animals fed NC in either our male (sham treatment: n = 8; irradiation treatment: n = 12) or female (sham treatment: n = 12; irradiation treatment: n = 12) cohorts (Fig. 3c). In contrast, HFD-fed female mice showed a significant reduction in weight gain following irradiation relative to sham controls, as previously reported (Lee et al. 2012). At 9 weeks post-treatment, irradiated female mice receiving HFD (n = 10) had a $32 \pm 4\%$ increase in weight gain relative to sham controls (n = 9), which showed a $52 \pm 6\%$ increase in weight gain (p-value = 0.028, Fig. 3c). In sharp contrast, no significant differences in weight gain were observed in irradiated HFD-fed males relative to sham controls ($59 \pm 7\%$ increase, n = 11, vs $71 \pm 7\%$ increase, n = 12, p = 0.27; Fig. 3c). These data confirm previous reports that neurons generated in the female ME in response to HFD act to promote energy storage (Lee et al. 2012, 2013).

Discussion

Several recent studies have reported that neurogenesis occurs in the adult hypothalamus (Migaud et al. 2010; Lee and Blackshaw 2012, 2014), a central regulator of metabolism and energy balance. We investigated how changes in diet can modulate hypothalamic neurogenesis by presenting young adult mice with different diets. We observed that HFD, LPD, and CR HFD all differentially modulated proliferation and neurogenesis in the hypothalamic ME and ArcN, hypothalamic regions that regulate energy balance (summarized in Fig. 4b). We observed an increase in ME neurogenesis and a decrease in ArcN neurogenesis in response to both HFD and LPD (Fig. 1c, d). In contrast, HFD selectively activated neurogenesis in the female ME. These differential region-specific changes in neurogenesis may lead to differential generation of orexinergic and anorexinergic neurons in response to dietary cues and may serve as a mechanism that allows adaptation to long-term changes in energy balance homeostasis.

These region-specific changes in adult hypothalamic neurogenesis most likely are mediated by differing exposures to secreted peptide, growth factors, and neurotrophic factors that signify feeding status and long-term energy availability. The ME, by virtue of its access to the third ventricle and status as a circumventricular organ, is exposed to a variety of these secreted factors via the CSF and the blood. By comparison, the ArcN, a structure protected by the blood–brain barrier (Mullier et al. 2010), has less access to circulating satiety signals and hormones. The permeability of hypothalamic blood–brain barrier to blood-borne factors has been proposed to be differentially regulated in fed and fasting conditions through a VEGF-dependent mechanism (Langlet et al. 2013a). Continuous integration of these peripheral signals by neurons belonging to both the ArcN and the ME of the hypothalamus is critical for central regulation of energy balance and neuro-endocrine function (Schaeffer et al. 2013). Our data suggest that adult-generated

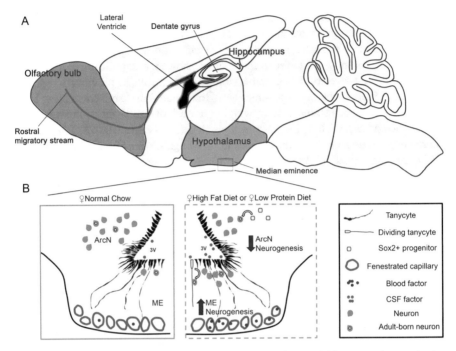

Fig. 4 Regulation of hypothalamic neurogenesis by dietary factors. (**a**) In mammals, constitutive adult neurogenesis is primarily confined to three brain regions (highlighted in *orange*). The hippocampal dentate gyrus and the subventricular zone of the lateral ventricles are canonical neurogenic niches. Additionally, recent observations demonstrate that the ventrobasal hypothalamus serves as a neurogenic niche, engaging in low but constitutive levels of neurogenesis in adults. (**b**) Hypothalamic tanycytes and parenchymal Sox2[+] cells represent potential hypothalamic neural progenitor populations that give rise to adult-born neurons. In the hypothalamic ME, tanycytes are a convincing neurogenic source for the ArcN and ME. In females, both HFD and low protein diet (LPD) increase ME neurogenesis in the ME, while concurrently decreasing ArcN neurogenesis in the ArcN. Several lines of evidence suggest that this dynamic change in neurogenesis is mediated between a tanycytic neural progenitor pool and factors present in the cerebrospinal fluid (CSF) and/or those circulating through the fenestrated capillaries of the ME

neurons in both hypothalamic regions may show differing sensitivities to dietary and hormonal signals that help maintain energy homeostasis.

Several secreted factors that signal feeding status and long-term energy availability regulate adult neurogenesis in various neurogenic niches, including the hypothalamus (Sousa-Ferreira et al. 2014). For instance, compared to NC-fed controls, HFD-fed mice exhibited substantially higher ciliary neurotrophic factor (CNTF) mRNA in tanycytes and multi-ciliated ependymal cells, whereas CR mice showed substantially lower expression levels; this coincided with similar changes in CNTF receptor (CNTFR) mRNA (Severi et al. 2013). Taken together with previous findings that CNTF delivered by i.c.v. cannulation stimulates adult hypothalamic neurogenesis (Kokoeva et al. 2005), this finding suggests that dietary signals may regulate hypothalamic neurogenesis in the ME through altered CNTF signaling. This hypotheses is supported by observations that beta2 tanycytes of the adult hypothalamic proliferative zone (HPZ) are enriched with CNTFR, as compared to alpha 1,2 tanycytes (Kokoeva et al. 2005). In that study, mice receiving intracerebroventricular infusion of CNTF demonstrated hyperplasia within the HPZ of the ME, as indicated by particularly high levels of BrdU incorporation (Kokoeva et al. 2005). The development of inducible Cre mouse lines specific for hypothalamic neural progenitors will help identify additional signaling pathways that are critical for the regulation of hypothalamic neurogenesis (Robins et al. 2013a, b; Pak et al. 2014).

In female mice fed a NC diet, we observed relatively low levels of hypothalamic neurogenesis in the ME. Interestingly, upon presentation of a dietary challenge such as HFD, ME neurogenesis was increased in females. Moreover, computer tomography-guided radiological inhibition of ME neurogenesis (Lee et al. 2013) reduced HFD-induced weight gain in females, but not in males (Fig. 3b). These intriguing results suggest that weight gain in females can be attributed in part to additional adult-generated neurons in the ME and that the neural circuitry regulating body weight differs in some respects between females and males. These findings are consistent with previous studies demonstrating that sex hormones can regulate hypothalamic neurogenesis in a region-dependent manner (Ahmed et al. 2008; Cheng 2013). Taken together with our results, this body of work highlights the importance of examining both the regional differences in hypothalamic neurogenesis and the sex-specific differences. Such differences likely at least partially account for differences in the levels and diet dependence of adult hypothalamic neurogenesis observed by different groups.

Factors that mediate these sex-dependent differences in hypothalamic neurogenesis have not yet been identified but could involve numerous levels of regulation, such as hormone-dependent plasticity (de Seranno et al. 2010), differences in blood-barrier access between the sexes (Hoxha et al. 2013), and hormone-specific induction of feeding behavior (Sieck et al. 1978). Lastly, it is possible that it is the survival of newborn neurons, rather than (or in addition to) their proliferation, that is sexually dimorphic, as has been previously demonstrated for prenatally generated hypothalamic neurons (Waters and Simerly 2009; Forger et al. 2004; Tobet and Hanna 1997; Park et al. 1998).

In addition to being a means of regulating energy homeostasis in response to long-term changes in diet, adult hypothalamic neurogenesis may be triggered in response to environmental injury. The hypothalamic ME, in contrast to other hypothalamic regions, lies outside of the blood–brain barrier and is thus directly exposed to circulating toxins and pathogens, as well as nutrients that can lead to cellular injury when in oversupply. Hypothalamic neural injury and inflammation are seen in obese animals and humans (Thaler et al. 2012, 2013; Li et al. 2012). The increased neurogenesis in adult female ME may serve to replace damaged neurons in this region. Indeed, at least one study has reported that neurons important for energy balance regulation can be regenerated in adult hypothalamus in response to neurodegenerative-like injury (Pierce and Xu 2010). Further studies to determine the role of environmental and physiological factors in regulating adult hypothalamic neurogenesis may yet reveal new mechanistic approaches towards the treatment of obesity and metabolic disorders.

Acknowledgments We would like to thank W. Yap, B. Clark and J. Bedont for their insightful comments on the manuscript. This work was supported by NIH R01DK108230 and a Baltimore Diabetes Research Center Training and Feasibility grant to S.B. S.B. was a W.M. Keck Distinguished Young Scholar in Medical Research.

References

Ahmed EI, Zehr JL, Schulz KM, Lorenz BH, DonCarlos LL, Sisk CL (2008) Pubertal hormones modulate the addition of new cells to sexually dimorphic brain regions. Nat Neurosci 11:995–997

Batailler M, Droguerre M, Baroncini M, Fontaine C, Prevot V, Migaud M (2013) DCX expressing cells in the vicinity of the hypothalamic neurogenic niche: a comparative study between mouse, sheep and human tissues. J Comp Neurol 522:1966–1985

Batailler M, Derouet L, Butruille L, Migaud M (2015) Sensitivity to the photoperiod and potential migratory features of neuroblasts in the adult sheep hypothalamus. Brain Struct Funct [Epub ahead of print]

Bennett L, Yang M, Enikolopov G, Iacovitti L (2009) Circumventricular organs: a novel site of neural stem cells in the adult brain. Mol Cell Neurosci 41:337–347

Bless EP, Reddy T, Acharya KD, Beltz BS, Tetel MJ (2014) Oestradiol and diet modulate energy homeostasis and hypothalamic neurogenesis in the adult female mouse. J Neuroendocrinol 26:805–816

Bolborea M, Dale N (2013) Hypothalamic tanycytes: potential roles in the control of feeding and energy balance. Trends Neurosci 36:91–100

Cai D (2013) Neuroinflammation and neurodegeneration in overnutrition-induced diseases. Trends Endocrinol Metab 24:40–47

Cheng MF (2013) Hypothalamic neurogenesis in the adult brain. Front Neuroendocrinol 34:167–178

Dahiya S, da Lee Y, Gutmann DH (2011) Comparative characterization of the human and mouse third ventricle germinal zones. J Neuropathol Exp Neurol 70:622–633

de Seranno S, d'Anglemont de Tassigny X, Estrella C, Loyens A, Kasparov S, Leroy D, Ojeda SR, Beauvillain JC, Prevot V (2010) Role of estradiol in the dynamic control of tanycyte plasticity mediated by vascular endothelial cells in the median eminence. Endocrinology 151:1760–1772

Ford EC, Achanta P, Purger D, Armour M, Reyes J, Fong J, Kleinberg L, Redmond K, Wong J, Jang MH, Jun H, Song HJ, Quinones-Hinojosa A (2011) Localized CT-guided irradiation inhibits neurogenesis in specific regions of the adult mouse brain. Radiat Res 175:774–783

Forger NG, Rosen GJ, Waters EM, Jacob D, Simerly RB, de Vries GJ (2004) Deletion of Bax eliminates sex differences in the mouse forebrain. Proc Natl Acad Sci U S A 101:13666–13671

Fry M, Hoyda TD, Ferguson AV (2007) Making sense of it: roles of the sensory circumventricular organs in feeding and regulation of energy homeostasis. Exp Biol Med (Maywood) 232:14–26

Gouaze A, Brenachot X, Rigault C, Krezymon A, Rauch C, Nedelec E, Lemoine GJ, Bauer S, Pénicaud L, Benani A (2013) Cerebral cell renewal in adult mice controls the onset of obesity. PLoS One 8, e72029

Haan N, Goodman T, Najdi-Samiei A, Stratford CM, Rice R, El Agha E, Bellusci S, Hajihosseini MK (2013) Fgf10-expressing tanycytes add new neurons to the appetite/energy-balance regulating centers of the postnatal and adult hypothalamus. J Neurosci 33:6170–6180

Hourai A, Miyata S (2013) Neurogenesis in the circumventricular organs of adult mouse brains. J Neurosci Res 91:757–770

Hoxha V, Lama C, Chang PL, Saurabh S, Patel N, Olate N, Dauwalder B (2013) Sex-specific signaling in the blood–brain barrier is required for male courtship in Drosophila. PLoS Genet 9, e1003217

Kisliouk T, Cramer T, Meiri N (2014) Heat stress attenuates new cell generation in the hypothalamus: a role for miR-138. Neuroscience 277:624–636

Kokoeva MV, Yin H, Flier JS (2005) Neurogenesis in the hypothalamus of adult mice: potential role in energy balance. Science 310:679–683

Kokoeva MV, Yin H, Flier JS (2007) Evidence for constitutive neural cell proliferation in the adult murine hypothalamus. J Comp Neurol 505:209–220

Langlet F, Levin BE, Luquet S, Mazzone M, Messina A, Dunn-Meynell AA, Balland Lacombe A, Mazur D, Carmeliet P, Bouret SG, Prevot V, Dehouck B (2013a) Tanycytic VEGF-A boosts blood-hypothalamus barrier plasticity and access of metabolic signals to the arcuate nucleus in response to fasting. Cell Metab 17:607–617

Langlet F, Mullier A, Bouret SG, Prevot V, Dehouck B (2013b) Tanycyte-like cells form a blood-cerebrospinal fluid barrier in the circumventricular organs of the mouse brain. J Comp Neurol 521:3389–3405

Lee DA, Blackshaw S (2012) Functional implications of hypothalamic neurogenesis in the adult mammalian brain. Int J Dev Neurosci 30:615–621

Lee DA, Blackshaw S (2014) Feed your head: neurodevelopmental control of feeding and metabolism. Annu Rev Physiol 76:197–223

Lee DA, Bedont JL, Pak T, Wang H, Song J, Miranda-Angulo A, Takiar V, Charubhumi V, Balordi F, Takebayashi H, Aja S, Ford E, Fishell G, Blackshaw S (2012) Tanycytes of the hypothalamic median eminence form a diet-responsive neurogenic niche. Nat Neurosci 15:700–702

Lee DA, Salvatierra J, Velarde E, Wong J, Ford EC, Blackshaw S (2013) Functional interrogation of adult hypothalamic neurogenesis with focal radiological inhibition. J Vis Exp (81):e50716

Li J, Tang Y, Cai D (2012) IKKbeta/NF-kappaB disrupts adult hypothalamic neural stem cells to mediate a neurodegenerative mechanism of dietary obesity and pre-diabetes. Nat Cell Biol 14:999–1012

Matsuzaki K, Katakura M, Inoue T, Hara T, Hashimoto M, Shido O (2015) Aging attenuates acquired heat tolerance and hypothalamic neurogenesis in rats. J Comp Neurol 523:1190–1201

McNay DE, Briancon N, Kokoeva MV, Maratos-Flier E, Flier JS (2012) Remodeling of the arcuate nucleus energy-balance circuit is inhibited in obese mice. J Clin Invest 122:142–152

McPherson AD, Barrios JP, Luks-Morgan SJ, Manfredi JP, Bonkowsky JL, Douglass AD, Dorsky RI (2016) Motor behavior mediated by continuously generated dopaminergic neurons in the zebrafish hypothalamus recovers after cell ablation. Curr Biol 26:263–269

Migaud M, Batailler M, Segura S, Duittoz A, Franceschini I, Pillon D (2010) Emerging new sites for adult neurogenesis in the mammalian brain: a comparative study between the hypothalamus and the classical neurogenic zones. Eur J Neurosci 32:2042–2052

Mohr MA, Sisk CL (2013) Pubertally born neurons and glia are functionally integrated into limbic and hypothalamic circuits of the male Syrian hamster. Proc Natl Acad Sci U S A 110:4792–4797

Mullier A, Bouret SG, Prevot V, Dehouck B (2010) Differential distribution of tight junction proteins suggests a role for tanycytes in blood-hypothalamus barrier regulation in the adult mouse brain. J Comp Neurol 518:943–962

Nogueira AB, Sogayar MC, Colquhoun A, Siqueira SA, Nogueira AB, Marchiori PE, Teixeira MJ (2014) Existence of a potential neurogenic system in the adult human brain. J Transl Med 12:75

Pak T, Yoo S, Miranda-Angulo LA, Blackshaw S (2014) Rax-CreER^{T2} knock-in mice: a tool for selective and conditional gene deletion in progenitors and radial glia of the retina and hypothalamus. PLoS One 9, e90381

Park JJ, Tobet SA, Baum MJ (1998) Cell death in the sexually dimorphic dorsal preoptic area/anterior hypothalamus of perinatal male and female ferrets. J Neurobiol 34:242–252

Pencea V, Bingaman KD, Wiegand SJ, Luskin MB (2001) Infusion of brain-derived neurotrophic factor into the lateral ventricle of the adult rat leads to new neurons in the parenchyma of the striatum, septum, thalamus, and hypothalamus. J Neurosci 21:6706–6717

Perez-Martin M, Cifuentes M, Grondona JM, Lopez-Avalos MD, Gomez-Pinedo U, Garcia-Verdugo JM, Fernandez-Llebrez P (2010) IGF-I stimulates neurogenesis in the hypothalamus of adult rats. Eur J Neurosci 31:1533–1548

Pierce AA, Xu AW (2010) De novo neurogenesis in adult hypothalamus as a compensatory mechanism to regulate energy balance. J Neurosci 30:723–730

Purkayastha S, Cai D (2013) Neuroinflammatory basis of metabolic syndrome. Mol Metab 2:356–363

Robins SC, Trudel E, Rotondi O, Liu X, Djogo T, Kryzskaya D, Bourque CW, Kokoeva MV (2013a) Evidence for NG2-glia derived, adult-born functional neurons in the hypothalamus. PLoS One 8, e78236

Robins SC, Stewart I, McNay DE, Taylor V, Giachino C, Goetz M, Ninkovic J, Briancon N, Maratos-Flier E, Flier JS, Kokoeva MV, Placzek M (2013a) Alpha-Tanycytes of the adult hypothalamic third ventricle include distinct populations of FGF-responsive neural progenitors. Nat Commun 4: 2049

Sachs BD, Caron MG (2014) Chronic fluoxetine increases extra-hippocampal neurogenesis in adult mice. Int J Neuropsychopharmacol 18(4). doi:10.1093/ijnp/pyu029

Schaeffer M, Langlet F, Lafont C, Molino F, Hodson DJ, Roux T, Lamarque L, Verdié P, Bourrier E, Dehouck B, Banères JL, Martinez J, Méry PF, Marie J, Trinquet E, Fehrentz JA, Prévot V, Mollard P (2013) Rapid sensing of circulating ghrelin by hypothalamic appetite-modifying neurons. Proc Natl Acad Sci U S A 110:1512–1517

Severi I, Perugini J, Mondini E, Smorlesi A, Frontini A, Cinti S, Giordano A (2013) Opposite effects of a high-fat diet and calorie restriction on ciliary neurotrophic factor signaling in the mouse hypothalamus. Front Neurosci 7:263

Sieck GC, Nance DM, Gorski RA (1978) Estrogen modification of feeding behavior in the female rat: influence of metabolic state. Physiol Behav 21:893–897

Sousa-Ferreira L, de Almeida LP, Cavadas C (2014) Role of hypothalamic neurogenesis in feeding regulation. Trends Endocrinol Metab 25:80–88

Thaler JP, Yi CX, Schur EA, Guyenet SJ, Hwang BH, Dietrich MO, Zhao X, Sarruf DA, Izgur V, Maravilla KR, Nguyen HT, Fischer JD, Matsen ME, Wisse BE, Morton GJ, Horvath TL, Baskin DG, Tschöp MH, Schwartz MW (2012) Obesity is associated with hypothalamic injury in rodents and humans. J Clin Invest 122:153–162

Thaler JP, Guyenet SJ, Dorfman MD, Wisse BE, Schwartz MW (2013) Hypothalamic inflammation: marker or mechanism of obesity pathogenesis? Diabetes 62:2629–2634

Tobet SA, Hanna IK (1997) Ontogeny of sex differences in the mammalian hypothalamus and preoptic area. Cell Mol Neurobiol 17:565–601

Wang X, Kopinke D, Lin J, McPherson AD, Duncan RN, Otsuna H, MoroE HK, Grunwald DJ, Argenton F, Chien CB, Murtaugh LC, Dorsky RI (2012) Wnt signaling regulates postembryonic hypothalamic progenitor differentiation. Dev Cell 23:624–636

Waters EM, Simerly RB (2009) Estrogen induces caspase-dependent cell death during hypothalamic development. J Neurosci 29:9714–9718

Xu Y, Tamamaki N, Noda T, Kimura K, Itokazu Y, Matsumoto N, Dezawa M, Ide C (2005) Neurogenesis in the ependymal layer of the adult rat 3rd ventricle. Exp Neurol 192:251–264

Genetic Dissection of the Neuroendocrine and Behavioral Responses to Stressful Challenges

Alon Chen

Abstract Dysregulation of the stress response is implicated in many psychopathologies. Data gathered over the past two decades have proposed a rather dualistic view of the central corticotropin-releasing factor (CRF)-urocortin system. Originally, it was thought that CRF/CRF receptor type 1 (CRFR1) signaling mediated stress-initiated effects and increased anxiety-like behavior, whereas activation of urocortins/CRFR2 ensured adequate recovery from stress and restoration of homeostasis. However, this view was based on data gained from genetically modified mouse models and pharmacological approaches; now, with the emergence of new and more specific biological tools, it has become clear that this is an over-simplistic proposal. It is becoming apparent that the function of the CRF-urocortin system's components relies profoundly on the spatial and temporal patterns of activity of the CRF family members. Here, we provide an overview of recent work that proposes a more dynamic, modulatory role for the CRF system's central pathways in the modulation of stress-linked behaviors. Recent findings suggest that the CRF system's actions are brain-region specific and dependent on the type of neuronal cell involved.

In 1955, Hans Seyle recognized that the hypothalamic-pituitary-adrenal (HPA) axis orchestrates the stress response (Selye 1955). However, it took until 1981 for the structure of its principal regulatory peptide, corticotropin-releasing factor (CRF), to be characterized (Vale et al. 1981). Neurons in the paraventricular nucleus (PVN) of the hypothalamus secrete CRF, which binds to receptors in the pituitary, which, in turn, secretes adrenocorticotropic hormone (ACTH) into the circulation. The circulating ACTH binds at the adrenal cortex stimulating the synthesis and secretion of glucocorticoids: cortisol in humans and corticosterone in rodents. Glucocorticoids are the downstream biological effector of the neuroendocrine stress response and also provide negative feedback in the HPA axis.

A. Chen (✉)
Department of Stress Neurobiology and Neurogenetics, Max Planck Institute of Psychiatry, Munich, Germany

Department of Neurobiology, Weizmann Institute of Science, Rehovot, Israel
e-mail: alon_chen@psych.mpg.de

© The Author(s) 2016
D. Pfaff, Y. Christen (eds.), *Stem Cells in Neuroendocrinology*, Research and Perspectives in Endocrine Interactions, DOI 10.1007/978-3-319-41603-8_6

69

The CRF family of peptides includes the 41 amino acid peptide CRF and the more recently discovered urocortin (UCN) 1, 2 and 3. As shown in Fig. 1, this family of peptides binds to two distinct receptors: the CRF receptor type 1 (CRFR1) and type 2 (CRFR2). CRF is a high-affinity ligand for the CRFR1, whereas UCN1 binds with equal affinity to both receptors. UCN2 and UCN3 preferentially bind the CRFR2. However, specificity seems to be lost at higher concentrations of the ligand, with CRF activating CRFR2. CRFR1 and CRFR2 are produced from distinct genes and have numerous splice variants (CRFR1α,β, and CRFR2α,β,γ), some of which are nonfunctional (Perrin and Vale 1999; Grammatopoulos and Chrousos 2002).

CRF receptors are G_s protein-coupled receptors. They stimulate a cascade of events, firstly adenylate cyclase, which subsequently activates cyclic AMP (cAMP). Cyclic AMP stimulates protein kinase A (PKA) to phosphorylate substrates such as the cAMP responsive-element–binding (CREB) protein, thereby inducing transcription of downstream target genes. In addition, cAMP binds to exchange proteins, which then activate the extracellular signal-regulated kinase-mitogen-activated protein kinase (ERK-MAPK) cascade (Gutknecht et al. 2009). The ERK/MAPK pathway regulates synaptic plasticity, including dendrite stabilization, ion channel transmission, transcription of CREB and other genes, and receptor scaffolding, trafficking, and crosstalk. Coupling to G_s is the dominant mechanism for stimulating intracellular calcium mobilization by CRFR1 and CRFR2. However, CRF receptors also interact with other G protein systems, including G_q, G_i, G_o, $G_{il/2}$, and G_z (Grammatopoulos et al. 2001). Coupling to G_q

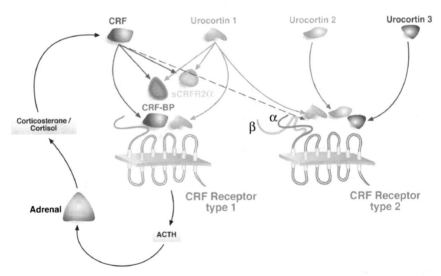

Fig. 1 The CRF family of neuropeptides and their action on the CRFR1 and CRFR2 (Kuperman and Chen 2008, with permission). *Colored arrows* indicate the receptors with which each ligand preferably interacts; *dashed lines* indicate a lower affinity binding. CRFR2 has two apparent membrane-bound splice variants in rodents, resulting in two receptor proteins, CRFR2α and CRFR2β

activates PLC$_\beta$, which cleaves the phospholipid phosphatidylinositol 4,5-bisphosphate into diacyl glycerol (DAG) and inositol 1,4,5-trisphosphate (IP$_3$). IP$_3$ then diffuses through the cytosol to bind to IP$_3$ receptors to increase intracellular calcium levels, whereas DAG activates protein kinase C (PKC), which activates ERK1/2 via Ras, Raf and MEKs. G$_{i/o}$ mediated activation of G$_{\beta\gamma}$ subunits activates the ERK1/2 by PI-3 kinase and induces PLCγ activation, which increases intracellular calcium release. Thus, depending on their localization and cellular context, CRF receptors are able to modulate a wide variety of signaling pathways (Hauger et al. 2006, 2009) and kinases [including PKA, protein kinase B (PKB), PKC (Gutknecht et al. 2009), mitogen activated protein (MAP) kinases (e.g., ERK1/2), and intracellular Ca^{2+} concentrations]. CRF receptors activate these various G-protein systems in a concentration-dependent manner (Grammatopoulos and Chrousos 2002).

Besides the activated pathway, the desensitization and internalization following receptor activation seems to depend on the specific ligand bound. Desensitization of CRFR2 cAMP signaling was shown to occur more rapidly and to a greater extent in response to UCN2 binding compared to UCN3, whereas CRF is a relatively weak desensitizing agonist (Gutknecht et al. 2008). Additionally, the internalization of CRFR2 was shown to be greater upon exposure to UCN2 (Markovic et al. 2008).

In situ hybridization histochemical studies (and recently generated reporter mice; Justice et al. 2008; Kuhne et al. 2012) have shown that CRFR1 is widely distributed throughout the brain. The highest expression levels are seen in neocortical, limbic, brain stem regions and the cerebellum (Fig. 2). Moderate levels are found in the dorsal and median raphe nuclei, and only low levels are found in the PVN of the hypothalamus (Chalmers et al. 1995; Van Pett et al. 2000). CRFR1 is expressed in diverse neuronal subpopulations. They have been found on cell bodies, dendritic shafts and dendritic spines of hippocampal neurons (Chen et al. 2004a,b, 2010). Compared to CRFR1, CRFR2 mRNA has a much more localized distribution pattern throughout the brain, which is virtually confined to subcortical structures. The highest levels of CRFR2 expression are found in the lateral septum, the

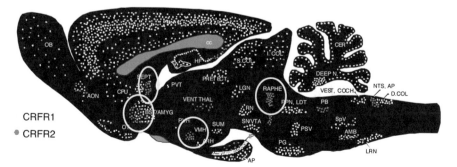

Fig. 2 The distribution of CRFR1 (*yellow*) and CRFR2 (*pink*) mRNA expression in the rodent brain. AMYG, amygdala; BST, bed nucleus of the stria terminalis; RAPHE, raphe nucleus; SEPT, lateral septum; VMH, ventromedial hypothalamic nucleus

ventromedial hypothalamic nucleus, and the choroid plexus (Fig. 2). Moderate levels are seen in the olfactory bulb, nuclei of the extended amygdala, hippocampus, the PVN and supraoptic nuclei of the hypothalamus, the inferior colliculus and the raphe nucleus (Chalmers et al. 1995; Van Pett et al. 2000). Only a limited amount of data is available on brain region-specific modulation of CRFR2 signaling. Among the most investigated regions are the extended amygdala, hippocampus, lateral septum, medial prefrontal cortex, and brain stem nuclei. Additionally, all four CRF family neuropeptides have been detected in the periphery, in particular UCN2 and UCN3, which have been recognized as novel modulators of centrally and peripherally controlled metabolic function (Chen et al. 2004a; Kuperman and Chen 2008).

Alterations in HPA axis function, attributed to centrally elevated CRF levels, were observed in depressed patients (Nemeroff et al. 1984; Holsboer 1999; Reul and Holsboer 2002) and in patients with post-traumatic stress disorder (PTSD; Bremner et al. 1997; McEwen 2002). These findings stimulated huge interest into the functioning of the CRF family. CRFR1 and CRFR2 seem to modulate anxiety-like behavior in a brain region-dependent manner. One obvious explanation for the differential behavioral effects of CRF receptor activation is that distinct brain regions simply serve distinct behavioral functions, and their activation would thus induce a distinct behavioral phenotype. Activation of brain regions implicated in emotional arousal, such as the extended amygdala, would be expected to increase anxiety, whereas activation of cognitive control regions, such as the medial prefrontal cortex, would contribute to stress coping and reduce anxiety. However, both the behavioral and electrophysiological effects of CRF receptor activation implicate a more complex underlying mechanism of region dependency and suggest specificity due to the cell type and ligand concentration.

Since specific CRF receptor antibodies are lacking, much about the action of the CRF family has been learned from transgenic mice, overexpressing or knocked out for the various members. Increased central levels of CRF [induced by either CRF administration (Britton et al. 1986; Dunn and File 1987; Dunn and Berridge 1990) or its overexpression in transgenic mice (Stenzel-Poore et al. 1994; Heinrichs et al. 1997; van Gaalen et al. 2002; Binneman et al. 2008)] produces anxiogenic behavior, whereas suppression induces anxiolytic effects (Skutella et al. 1994) and reduced stress-induced anxiety (Skutella et al. 1994). This action was attributed to CRFR1 activation specifically, since CRFR1 blockage by antisense treatment (Skutella et al. 1998; Liebsch et al. 1999) or selective antagonists (i.e., antalarmin; Habib et al. 2000; Valdez et al., 2002) prevented CRF/stress-induced anxiety. Furthermore, CRFR1 knock-out mice displayed reduced anxiety-like behavior (Smith et al. 1998; Timpl et al. 1998; Contarino et al. 1999; Muller et al. 2003). Taken together, these data suggested a critical role for CRFR1 activation in eliciting stress-induced anxiety. However, although these findings suggest a causative role for CRFR1 over-activation in stress-related psychopathologies, clinical trials of CRFR1 antagonists as potential next-generation anxiolytics/antidepressants have unfortunately met with little success.

In contrast to CRFR1, the role of CRFR2 in the mediation of psychopathologies is less clear, and two prominent theories are currently circulating that explain its role. The most popular view is that CRFR2 activation is responsible for ensuring homeostasis and counteracts the stress response-provoking effects and anxiety-like behavior induced by CRFR1 activation (Coste et al. 2001; Bale and Vale 2004; Heinrichs and Koob 2004; Muller and Holsboer 2006). This theory is primarily based on the increased corticosterone stress response observed in CRFR2 KO mice (Bale et al. 2000; Coste et al. 2000), and the anxiogenic phenotype observed in some (but not all; Coste et al., 2000) of the CRFR2 KO mouse lines (Bale et al. 2000; Kishimoto et al. 2000). However, the increased levels of anxiety and stress-related behaviors observed in a mouse line displaying (chronic) overexpression of UCN3 (Neufeld-Cohen et al. 2012), as well as the observed reduction in shock-induced freezing in response to reduced CRFR2 expression (by administration of CRFR2 mRNA antisense oligonucleotides; Ho et al. 2001) contradict this theory.

UCN1 neurons are mainly localized in the Edinger Westphal nucleus where they constitute the centrally projecting part of the nucleus to the lateral septum. UCN1 mRNA is also found in the lateral olivary and supraoptic nucleus (Bittencourt et al. 1999). UCN2 expression has been shown in several regions that are involved in the physiological and behavioral responses to stress, such as the PVN, locus coeruleus, and it partly overlaps with CRF and UCN1 expression in the hypothalamus and brainstem. UCN2 is thought to be the primary ligand for CRFR2s in the bed nucleus of the stria terminalis (BNST), PVN, central amygdala, parabrachial nucleus and nucleus tractus solitarii (Reyes et al. 2001). UCN3 also shows a distinct expression pattern and is mainly found in the medial preoptic area, rostral perifornical area, the posterior part of the BNST and the medial amygdala, with projections observed in the intermediate lateral septum (Lewis et al. 2001) and in the BNST (Deussing et al. 2010). However, although the sites of mRNA expression of these ligands are pretty well established, it is still largely unknown at which exact sites (i.e., axonal, dendritic, synaptic) the ligands are in fact released.

UCN1 neurons are recruited following chronic stress exposure and stay active for a prolonged period of time, suggesting that this peptide plays a prominent role in the later, adaptive phase of the stress response (Korosi et al. 2005; Xu et al. 2010; Ryabinin et al. 2012). The UCN family members' involvement in stress-related behavior was assessed in UCN1/UCN2 (dKO) double and UCN1/UCN2/UCN3 (tKO) triple knockout mice. Although the dKO knockout mice displayed no changes in basal HPA axis activity, they had elevated corticosterone levels following acute stress exposure (Neufeld-Cohen et al. 2010b). HPA axis function was unchanged in tKO mice compared to controls (Neufeld-Cohen et al. 2010a). The dKO knockout mice displayed decreased anxiety-like behavior under basal and acute stress conditions, which was accompanied by elevated serotonin concentrations in a number of brain regions, including the dorsal raphe nucleus, hippocampus, basolateral amygdala and subiculum (Neufeld-Cohen et al. 2010b). In contrast, tKO mice exhibited increased anxiety-like behavior, but only 24 h after restraint stress. Moreover, tKO mice displayed an increased stress-induced startle response

(Neufeld-Cohen et al. 2010b). As opposed to dKO mice, the behavioral phenotype in tKO mice was associated with decreased serotonergic metabolism in regions such as the septum, central and basolateral amygdala (Neufeld-Cohen et al. 2010a). Again, the effect of compensatory changes in CRF expression on emotional behavior cannot be excluded in many of the UCN mouse models, as shown in dKO mice (Neufeld-Cohen et al. 2010b). Overall, the data suggest that the UCN's binding to CRFR2 is able to regulate specific aspects of stress-related emotional behavior, complementing the effects of CRF to CRFR1.

Another great advance in the understanding of the CRF system's actions can be accredited to the specificity of genetic or viral-mediated loss-of-function approaches. For example, deletion of the CRFR1 in glutamatergic neurons reduces anxiety-related behavior, whereas deletion in dopaminergic neurons increases anxiety-related behavior (Refojo et al. 2011). Lentiviral-mediated knockdown of CRF in the central amygdala attenuated stress-induced anxiety-like behavior and altered HPA axis activity, reinforcing a role for amygdalar CRF in the modulation of fear and anxiety (Regev et al. 2011). Lentiviral knockdown of CRFR1 in the basolateral amygdala was also shown to decrease anxiety-like behavior and mimicked the anxiolytic effect of environmental enrichment (Sztainberg et al. 2010). Using stereotactic delivery of lentiviruses into the paraventricular nucleus, Elliott et al. (2010) targeted CRF by using lentiviral constructs that carried shRNA against CRF. This approach enabled the site-specific manipulation of gene expression. The knockdown attenuated chronic stress-induced social avoidance, which was shown to result from demethylation of the CRF promoter (Elliott et al. 2010).

The BNST is thought to be involved in the regulation of anxiety. It is positioned as a relay center between the limbic structures and mediates anticipatory stress. Lentiviral-mediated knockdown of CRFR2 in the BNST increased anxiety-related behavior both immediately and 24 h after restraint stress. Studying the knockdown and overexpression of CRFR2 specifically in the BNST revealed an important role of this receptor in PTSD-like behavior (Lebow et al. 2012).

In some of our most recent work, we have used optogenic regulation of posterior BNST CRFR2 neurons using bilateral fiber optic implantation. CRFR2 Cre mice were crossed with conditional channelrhodopsin (ChR) mice, in which blue light activates CRFR2-expressing neurons specifically, or with conditional halorhodopsin (NpHR2) mice, in which yellow light inhibits CRFR2-expressing neurons specifically. Using this tool, we found that specific activation of CRFR2 neurons in the posterior BNST reduced anxiety-related behavior and, conversely, their suppression increased anxiety-related behavior (Henckens et al. 2016). Similar viral-mediated loss-of-function tools were used to knockdown CRFR1 in the globus pallidus, revealing a previously unknown anxiolytic effect of the receptor in this brain region, which was further confirmed with site-specific CRFR1 antagonist administration (Sztainberg et al. 2011). The relevance of CRFR1 receptors in addiction and reward processes was investigated in the ventral tegmental area using lentiviral-mediated knockdown. Knockdown of CRFR1 reduced cue-induced and acute food deprivation, stress-induced cocaine seeking, but had no effect on self-administering behavior. CRFR1 signalling in the ventral tegmental area presents a target for convergent effects of both cue- and stress-induced cocaine-seeking pathways (Chen et al. 2014).

In summary, recent advances in methodological approaches have begun elucidating the site-specific actions of the CRF system. Most notably, they suggest that, via the CRFR2, UCNs might be the central system involved in coping with stress. Improved understanding of the CRF system holds the promise of better treatment of stress-related psychopathologies.

References

Bale TL, Vale WW (2004) CRF and CRF receptors: role in stress responsivity and other behaviors. Annu Rev Pharmacol Toxicol 44:525–557

Bale TL, Contarino A, Smith GW, Chan R, Gold LH, Sawchenko PE, Koob GF, Vale WW, Lee KF (2000) Mice deficient for corticotropin-releasing hormone receptor-2 display anxiety-like behaviour and are hypersensitive to stress. Nat Genet 24:410–414

Binneman B, Feltner D, Kolluri S, Shi Y, Qiu R, Stiger T (2008) A 6-week randomized, placebo-controlled trial of CP-316,311 (a selective CRH1 antagonist) in the treatment of major depression. Am J Psychiatry 165:617–620

Bittencourt JC, Vaughan J, Arias C, Rissman RA, Vale WW, Sawchenko PE (1999) Urocortin expression in rat brain: evidence against a pervasive relationship of urocortin-containing projections with targets bearing type 2 CRF receptors. J Comp Neurol 415:285–312

Bremner JD, Licinio J, Darnell A, Krystal JH, Owens MJ, Southwick SM, Nemeroff CB, Charney DS (1997) Elevated CSF corticotropin-releasing factor concentrations in posttraumatic stress disorder. Am J Psychiatry 154:624–629

Britton KT, Lee G, Vale W, Rivier J, Koob GF (1986) Corticotropin releasing factor (CRF) receptor antagonist blocks activating and 'anxiogenic' actions of CRF in the rat. Brain Res 369:303–306

Chalmers DT, Lovenberg TW, De Souza EB (1995) Localization of novel corticotropin-releasing factor receptor (CRF2) mRNA expression to specific subcortical nuclei in rat brain: comparison with CRF1 receptor mRNA expression. J Neurosci 15:6340–6350

Chen Y, Brunson KL, Adelmann G, Bender RA, Frotscher M, Baram TZ (2004a) Hippocampal corticotropin releasing hormone: pre- and postsynaptic location and release by stress. Neuroscience 126:533–540

Chen Y, Bender RA, Brunson KL, Pomper JK, Grigoriadis DE, Wurst W, Baram TZ (2004b) Modulation of dendritic differentiation by corticotropin-releasing factor in the developing hippocampus. Proc Natl Acad Sci U S A 101:15782–15787

Chen Y, Rex CS, Rice CJ, Dube CM, Gall CM, Lynch G, Baram TZ (2010) Correlated memory defects and hippocampal dendritic spine loss after acute stress involve corticotropin-releasing hormone signaling. Proc Natl Acad Sci U S A 107:13123–13128

Chen NA, Jupp B, Sztainberg Y, Lebow M, Brown RM, Kim JH, Chen A, Lawrence AJ (2014) Knockdown of CRF1 receptors in the ventral tegmental area attenuates cue- and acute food deprivation stress-induced cocaine seeking in mice. J Neurosci 34:11560–11570

Contarino A, Dellu F, Koob GF, Smith GW, Lee KF, Vale W, Gold LH (1999) Reduced anxiety-like and cognitive performance in mice lacking the corticotropin-releasing factor receptor 1. Brain Res 835:1–9

Coste SC, Kesterson RA, Heldwein KA, Stevens SL, Heard AD, Hollis JH, Murray SE, Hill JK, Pantely GA, Hohimer AR, Hatton DC, Phillips TJ, Finn DA, Low MJ, Rittenberg MB, Stenzel P, Stenzel-Poore MP (2000) Abnormal adaptations to stress and impaired cardiovascular function in mice lacking corticotropin-releasing hormone receptor-2. Nat Genet 24:403–409

Coste SC, Murray SE, Stenzel-Poore MP (2001) Animal models of CRH excess and CRH receptor deficiency display altered adaptations to stress. Peptides 22:733–741

Deussing JM, Breu J, Kuhne C, Kallnik M, Bunck M, Glasl L, Yen YC, Schmidt MV, Zurmuhlen R, Vogl AM, Gailus-Durner V, Fuchs H, Holter SM, Wotjak CT, Landgraf R, de Angelis MH, Holsboer F, Wurst W (2010) Urocortin 3 modulates social discrimination abilities via corticotropin-releasing hormone receptor type 2. J Neurosci 30:9103–9116

Dunn AJ, Berridge CW (1990) Is corticotropin-releasing factor a mediator of stress responses? Annu NY Acad Sci 579:183–191

Dunn AJ, File SE (1987) Corticotropin-releasing factor has an anxiogenic action in the social interaction test. Horm Behav 21:193–202

Elliott E, Ezra-Nevo G, Regev L, Neufeld-Cohen A, Chen A (2010) Resilience to social stress coincides with functional DNA methylation of the Crf gene in adult mice. Nat Neurosci 13:1351–1353

Grammatopoulos DK, Chrousos GP (2002) Functional characteristics of CRH receptors and potential clinical applications of CRH-receptor antagonists. Trends Endocrinol Metab 13:436–444

Grammatopoulos DK, Randeva HS, Levine MA, Kanellopoulou KA, Hillhouse EW (2001) Rat cerebral cortex corticotropin-releasing hormone receptors: evidence for receptor coupling to multiple G-proteins. J Neurochem 76:509–519

Gutknecht E, Hauger RL, Van der Linden I, Vauquelin G, Dautzenberg FM (2008) Expression, binding, and signaling properties of CRF2(a) receptors endogenously expressed in human retinoblastoma Y79 cells: passage-dependent regulation of functional receptors. J Neurochem 104:926–936

Gutknecht E, Van der Linden I, Van Kolen K, Verhoeven KF, Vauquelin G, Dautzenberg FM (2009) Molecular mechanisms of corticotropin-releasing factor receptor-induced calcium signaling. Mol Pharmacol 75:648–657

Habib KE, Weld KP, Rice KC, Pushkas J, Champoux M, Listwak S, Webster EL, Atkinson AJ, Schulkin J, Contoreggi C, Chrousos GP, McCann SM, Suomi SJ, Higley JD, Gold PW (2000) Oral administration of a corticotropin-releasing hormone receptor antagonist significantly attenuates behavioral, neuroendocrine, and autonomic responses to stress in primates. Proc Natl Acad Sci U S A 97:6079–6084

Hauger RL, Risbrough V, Brauns O, Dautzenberg FM (2006) Corticotropin releasing factor (CRF) receptor signaling in the central nervous system: new molecular targets. CNS Neurol Disord Drug Targets 5:453–479

Hauger RL, Risbrough V, Oakley RH, Olivares-Reyes JA, Dautzenberg FM (2009) Role of CRF receptor signaling in stress vulnerability, anxiety, and depression. Annu NY Acad Sci 1179:120–143

Heinrichs SC, Koob GF (2004) Corticotropin-releasing factor in brain: a role in activation, arousal, and affect regulation. J Pharmacol Exp Ther 311:427–440

Heinrichs SC, Min H, Tamraz S, Carmouche M, Boehme SA, Vale WW (1997) Anti-sexual and anxiogenic behavioral consequences of corticotropin-releasing factor overexpression are centrally mediated. Psychoneuroendocrinology 22:215–224

Henckens MJAG, Printz Y, Shamgar U, Lebow M, Drori Y, Kuehne C, Kolarz A, Deussing JM, Justice NJ, Yizhar O, Chen A (2016) The posterior bed nucleus of the stria terminalis critically contributes to stress recovery. Mol Psychiatry (in press)

Ho SP, Takahashi LK, Livanov V, Spencer K, Lesher T, Maciag C, Smith MA, Rohrbach KW, Hartig PR, Arneric SP (2001) Attenuation of fear conditioning by antisense inhibition of brain corticotropin releasing factor-2 receptor. Brain Res Mol Brain Res 89:29–40

Holsboer F (1999) The rationale for corticotropin-releasing hormone receptor (CRH-R) antagonists to treat depression and anxiety. J Psychiatr Res 33:181–214

Justice NJ, Yuan ZF, Sawchenko PE, Vale W (2008) Type 1 corticotropin-releasing factor receptor expression reported in BAC transgenic mice: implications for reconciling ligand-receptor mismatch in the central corticotropin-releasing factor system. J Comp Neurol 511:479–496

Kishimoto T, Radulovic J, Radulovic M, Lin CR, Schrick C, Hooshmand F, Hermanson O, Rosenfeld MG, Spiess J (2000) Deletion of crhr2 reveals an anxiolytic role for corticotropin-releasing hormone receptor-2. Nat Genet 24:415–419

Korosi A, Schotanus S, Olivier B, Roubos EW, Kozicz T (2005) Chronic ether stress-induced response of urocortin 1 neurons in the Edinger-Westphal nucleus in the mouse. Brain Res 1046:172–179

Kuhne C, Puk O, Graw J, Hrabe de Angelis M, Schutz G, Wurst W, Deussing JM (2012) Visualizing corticotropin-releasing hormone receptor type 1 expression and neuronal connectivities in the mouse using a novel multifunctional allele. J Comp Neurol 520:3150–3180

Kuperman Y, Chen A (2008) Urocortins: emerging metabolic and energy homeostasis perspectives. Trends Endocrinol Metab 19:122–129

Lebow M, Neufeld-Cohen A, Kuperman Y, Tsoory M, Gil S, Chen A (2012) Susceptibility to PTSD-like behavior is mediated by corticotropin-releasing factor receptor type 2 levels in the bed nucleus of the stria terminalis. J Neurosci 32:6906–6916

Lewis K, Li C, Perrin MH, Blount A, Kunitake K, Donaldson C, Vaughan J, Reyes TM, Gulyas J, Fischer W, Bilezikjian L, Rivier J, Sawchenko PE, Vale WW (2001) Identification of urocortin III, an additional member of the corticotropin-releasing factor (CRF) family with high affinity for the CRF2 receptor. Proc Natl Acad Sci U S A 98:7570–7575

Liebsch G, Landgraf R, Engelmann M, Lorscher P, Holsboer F (1999) Differential behavioural effects of chronic infusion of CRH 1 and CRH 2 receptor antisense oligonucleotides into the rat brain. J Psychiatr Res 33:153–163

Markovic D, Punn A, Lehnert H, Grammatopoulos DK (2008) Intracellular mechanisms regulating corticotropin-releasing hormone receptor-2beta endocytosis and interaction with extracellularly regulated kinase 1/2 and p38 mitogen-activated protein kinase signaling cascades. Mol Endocrinol 22:689–706

McEwen BS (2002) The neurobiology and neuroendocrinology of stress. Implications for post-traumatic stress disorder from a basic science perspective. Psychiatr Clin North Am 25 (469-494):ix

Muller MB, Holsboer F (2006) Mice with mutations in the HPA-system as models for symptoms of depression. Biol Psychiatry 59:1104–1115

Muller MB, Zimmermann S, Sillaber I, Hagemeyer TP, Deussing JM, Timpl P, Kormann MS, Droste SK, Kuhn R, Reul JM, Holsboer F, Wurst W (2003) Limbic corticotropin-releasing hormone receptor 1 mediates anxiety-related behavior and hormonal adaptation to stress. Nat Neurosci 6:1100–1107

Nemeroff CB, Widerlov E, Bissette G, Walleus H, Karlsson I, Eklund K, Kilts CD, Loosen PT, Vale W (1984) Elevated concentrations of CSF corticotropin-releasing factor-like immunoreactivity in depressed patients. Science 226:1342–1344

Neufeld-Cohen A, Tsoory MM, Evans AK, Getselter D, Gil S, Lowry CA, Vale WW, Chen A
    (2010a) A triple urocortin knockout mouse model reveals an essential role for urocortins in
    stress recovery. Proc Natl Acad Sci U S A 107:19020–19025
Neufeld-Cohen A, Evans AK, Getselter D, Spyroglou A, Hill A, Gil S, Tsoory M, Beuschlein F,
    Lowry CA, Vale W, Chen A (2010b) Urocortin-1 and -2 double-deficient mice show robust
    anxiolytic phenotype and modified serotonergic activity in anxiety circuits. Mol Psychiatry 15
    (426-441):339
Neufeld-Cohen A, Kelly PA, Paul ED, Carter RN, Skinner E, Olverman HJ, Vaughan JM, Issler O,
    Kuperman Y, Lowry CA, Vale WW, Seckl JR, Chen A, Jamieson PM (2012) Chronic
    activation of corticotropin-releasing factor type 2 receptors reveals a key role for 5-HT1A
    receptor responsiveness in mediating behavioral and serotonergic responses to stressful chal-
    lenge. Biol Psychiatry 72:437–447
Perrin MH, Vale WW (1999) Corticotropin releasing factor receptors and their ligand family.
    Annu NY Acad Sci 885:312–328
Refojo D, Schweizer M, Kuehne C, Ehrenberg S, Thoeringer C, Vogl AM, Dedic N,
    Schumacher M, von Wolff G, Avrabos C, Touma C, Engblom D, Schutz G, Nave KA,
    Eder M, Wotjak CT, Sillaber I, Holsboer F, Wurst W, Deussing JM (2011) Glutamatergic
    and dopaminergic neurons mediate anxiogenic and anxiolytic effects of CRHR1. Science
    333:1903–1907
Regev L, Neufeld-Cohen A, Tsoory M, Kuperman Y, Getselter D, Gil S, Chen A (2011) Prolonged
    and site-specific over-expression of corticotropin-releasing factor reveals differential roles for
    extended amygdala nuclei in emotional regulation. Mol Psychiatry 16:714–728
Reul JM, Holsboer F (2002) Corticotropin-releasing factor receptors 1 and 2 in anxiety and
    depression. Curr Opin Pharmacol 2:23–33
Reyes TM, Lewis K, Perrin MH, Kunitake KS, Vaughan J, Arias CA, Hogenesch JB, Gulyas J,
    Rivier J, Vale WW, Sawchenko PE (2001) Urocortin II: a member of the corticotropin-
    releasing factor (CRF) neuropeptide family that is selectively bound by type 2 CRF receptors.
    Proc Natl Acad Sci U S A 98:2843–2848
Ryabinin AE, Tsoory MM, Kozicz T, Thiele TE, Neufeld-Cohen A, Chen A, Lowery-Gionta EG,
    Giardino WJ, Kaur S (2012) Urocortins: CRF's siblings and their potential role in anxiety,
    depression and alcohol drinking behavior. Alcohol 46:349–357
Selye H (1955) Stress and disease. Science 122:625–631
Skutella T, Criswell H, Moy S, Probst JC, Breese GR, Jirikowski GF, Holsboer F (1994)
    Corticotropin-releasing hormone (CRH) antisense oligodeoxynucleotide induces anxiolytic
    effects in rat. Neuroreport 5:2181–2185
Skutella T, Probst JC, Renner U, Holsboer F, Behl C (1998) Corticotropin-releasing hormone
    receptor (type I) antisense targeting reduces anxiety. Neuroscience 85:795–805
Smith GW, Aubry JM, Dellu F, Contarino A, Bilezikjian LM, Gold LH, Chen R, Marchuk Y,
    Hauser C, Bentley CA, Sawchenko PE, Koob GF, Vale W, Lee KF (1998) Corticotropin
    releasing factor receptor 1-deficient mice display decreased anxiety, impaired stress response,
    and aberrant neuroendocrine development. Neuron 20:1093–1102
Stenzel-Poore MP, Heinrichs SC, Rivest S, Koob GF, Vale WW (1994) Overproduction of
    corticotropin-releasing factor in transgenic mice: a genetic model of anxiogenic behavior. J
    Neurosci 14:2579–2584
Sztainberg Y, Kuperman Y, Tsoory M, Lebow M, Chen A (2010) The anxiolytic effect of
    environmental enrichment is mediated via amygdalar CRF receptor type 1. Mol Psychiatry
    15:905–917
Sztainberg Y, Kuperman Y, Justice N, Chen A (2011) An anxiolytic role for CRF receptor type
    1 in the globus pallidus. J Neurosci 31:17416–17424
Timpl P, Spanagel R, Sillaber I, Kresse A, Reul JM, Stalla GK, Blanquet V, Steckler T,
    Holsboer F, Wurst W (1998) Impaired stress response and reduced anxiety in mice lacking a
    functional corticotropin-releasing hormone receptor 1. Nat Genet 19:162–166

Valdez GR, Inoue K, Koob GF, Rivier J, Vale W, Zorrilla EP (2002) Human urocortin II: mild locomotor suppressive and delayed anxiolytic-like effects of a novel corticotropin-releasing factor related peptide. Brain Res 943:142–150

Vale W, Spiess J, Rivier C, Rivier J (1981) Characterization of a 41-residue ovine hypothalamic peptide that stimulates secretion of corticotropin and beta-endorphin. Science 213:1394–1397

van Gaalen MM, Stenzel-Poore MP, Holsboer F, Steckler T (2002) Effects of transgenic overproduction of CRH on anxiety-like behaviour. Eur J Neurosci 15:2007–2015

Van Pett K, Viau V, Bittencourt JC, Chan RK, Li HY, Arias C, Prins GS, Perrin M, Vale W, Sawchenko PE (2000) Distribution of mRNAs encoding CRF receptors in brain and pituitary of rat and mouse. J Comp Neurol 428:191–212

Xu L, Bloem B, Gaszner B, Roubos EW, Kozicz T (2010) Stress-related changes in the activity of cocaineand amphetamine-regulated transcript and nesfatin neurons in the midbrain non-preganglionic Edinger-Westphal nucleus in the rat. Neuroscience 170:478–488

Pituitary Stem Cells: Quest for Hidden Functions

Hugo Vankelecom

Abstract The pituitary is the core endocrine gland, ruling fundamental processes of body growth, metabolism, reproduction and stress. Over the past decade, it has progressively become clear that the pituitary, like many adult tissues, harbors a population of stem cells. While the molecular depiction of these cells is constantly expanding, their function remains essentially hidden. From recent studies, the picture is developing that the stem cells of the adult pituitary are highly quiescent and mainly come into play during pathological conditions.

Upon transgenic cell-ablation damage in the pituitary, the stem cell compartment is promptly turned on with expansion and expression of the missing hormone. This activation is accompanied by substantial regeneration of the lost hormonal cells, a restorative competence that was unexpected in the mature gland. This regenerative skill, however, rapidly disappears with aging, together with a decline in the number and fitness of the stem cells. One function of the adult pituitary stem cells may thus be hidden in the regenerative toolbox of the gland, at least during a specified and limited time window.

Recent work also showed activation of the pituitary stem cell compartment during tumor formation in the (mouse) gland. Moreover, pituitary tumors (from patients and mice) contain a candidate 'tumor stem cell' (TSC) population. The pathogenetic steps of initiation, expansion, invasion and recurrence of pituitary tumors remain far from understood. A link between the tumor-driving TSC and the pituitary stem cells may shed new light on this tumorigenic darkness.

To conclude, decoding the hidden functions of pituitary stem cells will not only lead to better fundamental insights into their role but may also expose (novel) targets for treating pituitary tumors and for regenerative intervention in pituitary deficiency, as caused by damage, tumors or aging. Yet, the journey in the 'hidden valley' of pituitary stem cell functions has only just begun, and a long distance still has to be walked.

H. Vankelecom (✉)
Department of Development and Regeneration, Cluster of Stem Cell Biology and Embryology, Unit of Stem Cell Research, KU Leuven (University of Leuven), Campus Gasthuisberg O&N4, Herestraat 49, 3000 Leuven, Belgium
e-mail: Hugo.Vankelecom@kuleuven.be

D. Pfaff, Y. Christen (eds.), *Stem Cells in Neuroendocrinology*, Research and Perspectives in Endocrine Interactions, DOI 10.1007/978-3-319-41603-8_7

81

Introduction

The pituitary gland, in unison with the hypothalamus, constitutes the hub of our endocrine system, governing fundamental processes of growth, metabolism, sexual development, procreation and coping with immune and stress challenges (Melmed 2010; Vankelecom 2012; Willems and Vankelecom 2014). Due to this strategic position, malfunctioning of the pituitary leads to important morbidities that can be life-threatening (Schneider et al. 2007; Willems and Vankelecom 2014). In the past decade, mounting evidence has been presented that the endocrine gland contains a population of stem cells, purportedly sitting there to deal with renewal of cells that are worn out or have been damaged (Vankelecom 2012; Vankelecom and Chen 2014). However, like in comparable adult tissues that do not turn over very actively such as the brain and lung (Alvarez-Buylla and Lim 2004; Rando 2006; Slack 2008; Vankelecom 2012; Wabik and Jones 2015), the function of the stem cells residing in the pituitary remains enigmatic. Compared to the expanding molecular stripping of the pituitary stem cell phenotype, functional characterization of the cells clearly lags behind. In general, stem cells in 'lazy' low-turnover tissues are highly quiescent, and activation only clearly emerges in conditions of disease or damage (Barker et al. 2010; Huch et al. 2013a,b; Rando 2006; Slack 2008; Vankelecom 2012; Vankelecom and Chen 2014; Wabik and Jones 2015). Regarding the yet hidden functions of the pituitary stem cells, recent studies have shed some light on this obscure domain, particularly in the context of early postnatal maturation, damage repair and tumor growth. In this review, recent findings are summarized and emerging views presented. In addition, some brief perspectives are offered regarding therapeutic implementation of this new knowledge.

Primer on Pituitary Biology and Pathology

The pituitary gland consists of the anterior pituitary (AP), the posterior pituitary (PP) and the intermediate lobe (IL; Fig. 1), the latter being only rudimentary in humans (Melmed 2010; Vankelecom 2010, 2012). The AP represents the major endocrine segment of the gland, containing different cell types that each produce (a) specific hormone(s). Growth hormone (GH) is produced by the somatotropes, prolactin (PRL) by lactotropes, adrenocorticotropic hormone (ACTH) by corticotropes, thyroid-stimulating hormone (TSH) by thyrotropes, and luteinizing hormone (LH) and/or follicle-stimulating hormone (FSH) by gonadotropes (Fig. 1). Hormone production by the AP is regulated by signals from the hypothalamus, such as growth hormone-releasing hormone (GHRH) and somatostatin, which stimulate and inhibit, respectively, pituitary GH synthesis and secretion. The pituitary hormones act on distal target organs where they mainly control the production of peripheral hormones (such as glucocorticoids from the adrenal cortex and estradiol or testosterone from the gonads), which in turn negatively feed back on the

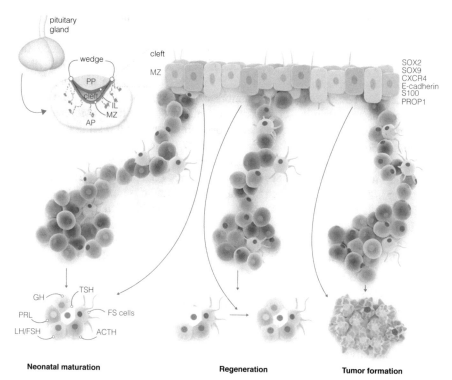

Fig. 1 Proposed model of pituitary stem cells and their functions. A cross-section of the (mouse) pituitary gland (*upper left*) shows the AP stem cell niche around the cleft (MZ; *red*) with the projected germinative wedge regions, and the putative secondary niches distributed all over the AP parenchyma (*red clusters*). The MZ stem cells and the in-gland stem cell clusters appear to be wired into a 3D network (see *dotted lines* in section and zoom-in drawing). The total stem cell pool is heterogeneous, encompassing stem (and further-advanced progenitor) cells in different phases of cell cycle, activation and maturation (indicated by *different shapes* and *red tints*), supposedly typified by different fingerprints of markers (of which the most dominant ones are indicated). To contribute to the pituitary endocrine cells, the stem/progenitor cells may move from the various niches into the glandular area (*arrows*), and EMT may be involved in this migration process. The pituitary stem cells are only modestly implicated in the low-rate turnover of the gland (see text) but appear to primarily pop up in the actively remodeling pituitary, as occurring during neonatal maturation (*left*), regeneration after cell/tissue destruction (*middle*), and development of tumors (with candidate TSC; *red*) (*right*). *AP* anterior pituitary, *FS* folliculo-stellate, *IL* intermediate lobe, *MZ* marginal zone, *PP* posterior pituitary

hypothalamus and pituitary. This multilayered interplay between hypothalamus, pituitary and target organs eventually generates tightly controlled, pulsatile secretion of hormones by the pituitary.

Because of its master position, disturbed function of the pituitary causes severe clinical complications (Schneider et al. 2007; Willems and Vankelecom 2014). For instance, GH deficiency causes growth disturbances and mental retardation in children and distorted fat metabolism, weakened muscles, osteoporosis and cardiovascular anomalies in adults, together strongly diminishing quality-of-life and life expectancy (Gasco et al. 2013). Pituitary deficiency (generally referred to as hypopituitarism) may be congenital or acquired during life, the latter accounting for the majority of cases (Mehta and Dattani 2008; Schneider et al., 2007; Willems and Vankelecom 2014; Zhu et al. 2005). Tumor growth in the pituitary (with a prevalence for clinically observed tumors of 1:1000) can compress the healthy endocrine tissue, thereby compromising pituitary function (Melmed 2011; Vankelecom 2012). Also the treatment procedures of surgical resection or irradiation inflict damage on the pituitary, resulting in hypofunction. In addition, traumatic brain injury, ranging from car accidents to sport-related impacts (e.g.,by boxing), represents an important cause of pituitary deficiency (Gasco et al. 2012; Tanriverdi et al. 2015). The current approach to treat hypopituitarism symptoms is by hormone replacement therapy (HRT) (Gasco et al. 2013; Schneider et al. 2007; Van Aken and Lamberts 2005; Willems and Vankelecom 2014). However, the exogenous hormones also cause side effects, further adding to the reduced quality-of-life. Moreover, a fundamental shortcoming of HRT lies in its failure to mimic the natural pulsatility of hormone release by the pituitary. Therefore, restoring deficient pituitary tissue and function would represent a more ideal tactic (Willems and Vankelecom 2014). Interestingly, it has recently been shown that the pituitary gland has the capacity to regenerate cells after destruction (Fu and Vankelecom 2012; Fu et al. 2012). Moreover, the local stem cells appear to participate and function as activated restorers (see below). This new finding may open the door toward regenerative opportunities.

The process of tumorigenesis in the pituitary is largely uncomprehended, and treatment remains suboptimal in up to two-thirds of patients (Melmed, 2003, 2011). A better understanding of the underlying mechanisms is essential to improve clinical management. In particular, the question arises whether pituitary tumors contain a driver population of 'tumor stem cells' (TSC), as has been discovered in other types of tumors (Clevers 2011; Dalerba et al. 2007; Sergeant et al. 2009; Wouters et al. 2009). Generally speaking, TCS are considered to power the development, growth, local invasion, metastasis, therapy resistance and/or recurrence of tumors. TSC thus represent a very appealing target to understand, treat and eradicate the tumor.

Pituitary Stem Cells: Expanding Molecular Portrayal

Stem cells have now been identified in many adult organs. They are considered the master builders of the tissue, replacing cells that have grown old and finished their task (homeostatic turnover), producing surplus cells when needed, or regenerating cells that were destroyed by physical or pathological insults (Alvarez-Buylla and Lim 2004; Rando 2006; Slack 2008; Vankelecom 2012; Vankelecom and Chen 2014; Wabik and Jones 2015). The tissue stem cells not only give rise to new specialized cells of the organ but must also perpetuate by producing at least one 'selfie.' Differentiation potential (ranging from uni- to multipotency) and self-renewal capacity indeed represent the core hallmarks of adult tissue stem cells (Barker et al. 2010; De Los Angeles et al. 2015; Rando 2006; Wabik and Jones 2015). On the other hand, the degree of activity of adult stem cells is clearly dependent on the tissue. Some organs show rapid cell turnover within days and contain highly energetic stem cells that drive this renewal (as, for instance, intestine and skin). Other organs are more static in their cell composition and only need cell replacement at a very low pace (such as the brain, heart, liver, pancreas, lung; Alvarez-Buylla and Lim 2004; Barker et al. 2010; Huch et al. 2013a,b; Rando 2006; Slack 2008; Vankelecom 2012; Wabik and Jones 2015). The pituitary belongs to the latter group, undergoing turnover in terms of months rather than days (Levy 2002, 2008; Nolan et al. 1998; Rando 2006; Vankelecom 2012; Vankelecom and Chen 2014). Stem cells in organs with low turnover appear highly dormant (quiescent) and are only jolted awake by strong triggers like loss of cells by damage. Stem cells, together with their further progressed progenitor cells (that have set off for differentiation), are typically housed in a dedicated setting of the tissue, the niche, containing cellular and molecular 'housemates' that control the maintenance, self-renewal and differentiation of the stem cell inhabitants (Alvarez-Buylla and Lim 2004; Rando 2006; Roskams 2006; Slack 2008; Vankelecom 2012; Vankelecom and Chen 2014; Yin et al. 2013).

The discovery of stem/progenitor cells in the pituitary was launched a decade ago by the identification of a side population (SP) in the adult gland of rodents and chicken (Chen et al. 2005). SP cells exhibit high efflux capacity, a property considered characteristic and important for stem cells to defend themselves against toxic substances. Meanwhile, the existence of stem cells in the pituitary has been clearly established (Chen et al. 2009; Fauquier et al. 2008; Garcia-Lavandeira et al. 2009). Their molecular portrayal is steadily expanding; a compact overview is provided here (for extensive reviews, see Vankelecom 2010, 2012; Vankelecom and Chen 2014).

Like stem cells of other tissues, pituitary stem cells in general express two classes of markers, i.e., broad stemness markers and pituitary embryogenesis-associated factors. Within the first group, the key stem-cell regulatory transcription factors SOX2 [SRY (sex determining region Y)-box 2] and SOX9 (Arnold et al. 2011; Jayakody et al. 2012; Pevny and Nicolis 2010) occupy a well-recognized position. The SOX factors 'color' cells that reside in the marginal zone (MZ), which borders the cleft (Fig. 1), a lumen that is left from the early-embryonic pituitary structure known as Rathke's pouch (RP; Vankelecom 2012). In addition, SOX2-immunoreactive (SOX2$^+$) cells are present as scattered clusters in the AP parenchymal area (Fig. 1; Chen et al. 2005, 2009; Fauquier et al. 2008; Fu et al. 2012; Garcia-Lavandeira et al. 2009; Gremeaux et al. 2012). The several locations of SOX2$^+$ cells may point to the existence of multiple stem cell niches in the pituitary, encompassing a major (primary) niche around the cleft and many secondary parenchymal niches, together taking care of dynamic and/or subtle cell adaptations in the gland. Interestingly, the SOX2$^+$ cell clusters and MZ cells appear to be connected (Fu et al. 2012; Gremeaux et al. 2012; Mollard et al. 2012;

Vankelecom and Chen 2014), thereby forming a three-dimensional (3D) network throughout the gland (Fig. 1), as also occurs among the different hormonal cell types (Le Tissier et al. 2012; Mollard et al. 2012). Such an integrated circuit would allow communication between the stem cells throughout the gland and promote coordinated actions. Of note, the pituitary stem cells seem to be part of the formerly identified heterogeneous population of folliculo-stellate (FS) cells within the gland (Allaerts and Vankelecom 2005), since the stem cells also express the FS cell marker S100 (Fig. 1). Thus, the SOX2$^+$ cell network may be part of the FS cell network, the first interconnected functional circuitry identified in the pituitary (Fauquier et al. 2001). In addition, the cell adhesion molecule E-cadherin (CDH1) is strongly expressed in the MZ, as well as in S100$^+$ cell patches spread over the AP lobe (Fauquier et al. 2008), and may participate in stitching the stem cells together (Fig. 1). Additional general stemness markers found in pituitary stem cells, although sometimes less well-defined, include OCT4, NANOG, glial cell line-derived neurotrophic factor receptor alpha 2 (GFRA2), NESTIN and Krüppel-like factor 4 (KLF4; reviewed in Vankelecom and Chen 2014). Of interest, SOX2, OCT4, NANOG and KLF4 constitute the transcription-factor core of embryonic stem (ES) and induced pluripotent stem (iPS) cells, the prototypical stem cells (De Los Angeles et al. 2015; Hyslop et al. 2005; Takahashi and Yamanaka 2006). Finally, the chemotactic receptor CXCR4 (C-X-C chemokine receptor type 4), which is more and more classified as a general stemness marker, is also upregulated in the pituitary stem cell compartment (Horiguchi et al. 2012; Vankelecom 2010, 2012; Vankelecom and Chen 2014).

Within the group of embryonic pituitary-related factors, expression of PROP1 (Prophet of Pit1) seems to be best defined. This transcription factor, which – among others - is essential for the regulation and migration of RP progenitor cells during pituitary embryogenesis (Himes and Raetzman 2009; Ward et al. 2005; Zhu et al. 2005, 2007), is expressed in the SOX2$^+$ cells, although it is not clear yet whether postnatal expression in the MZ stops after the first postnatal weeks or is continuous throughout life (Garcia-Lavandeira et al. 2009; Yoshida et al. 2009, 2011). In general, there is no co-localization of PROP1 or SOX2 with hormones (Chen et al. 2009; Fauquier et al. 2008; Fu et al. 2012a,b; Garcia-Lavandeira et al. 2009; Gremeaux et al. 2012; Yoshida et al. 2009), supporting the general notion that stem cell conservators must be downregulated before differentiation starts. The NOTCH pathway is another essential regulator of pituitary embryogenesis (Kita et al. 2007; Monahan et al. 2009; Raetzman et al. 2004; Zhu et al. 2005, 2007), of which several components are upregulated in the adult (mouse) pituitary stem cell fraction (Chen et al. 2006, 2009; Vankelecom 2010) and are found expressed in some of the MZ and parenchymal S100$^+$ cells of the (rat) AP (Tando et al. 2013). Some other transcriptional regulators that play an important role in pituitary embryonic development (such as *Hesx1*, the earliest known gene expressed in the pituitary primordium, *Lhx4*, *Pax6*, *Otx2*, *Ascl1* and the *Six/Eya/Dach* genes; Kelberman et al. 2009; Zhu et al. 2005, 2007) are also higher transcribed in the adult pituitary stem cell fraction (Chen et al. 2006, 2009; reviewed in Vankelecom 2010, 2012), as well as the cyclin-dependent kinase inhibitor *p57* that, during embryogenesis, emerges in the RP progenitor cells that stop cycling to embark on differentiation (Bilodeau et al., 2009; Vankelecom 2010; Vankelecom and Chen 2014). Finally, components of the fibroblast growth factor (FGF) and bone morphogenetic protein (BMP) pathways, having important roles in the specification and proliferation of RP progenitor cells (Ericson et al. 1998; Zhu et al. 2005, 2007), are also upregulated in the adult stem cells (Vankelecom 2010; Vankelecom and Chen 2014). Together, this expression portrait suggests that the stem cells of the adult pituitary (re-)use the embryonic developmental programs for their maintenance and progression toward differentiated cells (as also found in other tissues; Alvarez-Buylla and Lim 2004; Jensen et al. 2005; Roskams 2006; Slack 2008; Wagers and Conboy 2005).

Finally, a battery of new candidate markers has emerged from recent studies (extensively reviewed in Vankelecom 2010 and Vankelecom and Chen 2014), including the retinoic acid-producing retinal aldehyde dehydrogenase 1 (*Raldh1*; Fujiwara et al. 2007) and the retinoic acid receptor *Rarb* (Vankelecom 2010; Vankelecom and Chen 2014); the transcription factors PRRX1 and PRRX2 (expressed in proliferating RP progenitor cells until cell-cycle exit and start of differentiation; Susa et al. 2012); the cell-adhesion molecule cadherin-18 (expressed in the MZ in a different pattern than CDH1; Chauvet et al. 2009); the coxsackievirus and adenovirus receptor (CAR; Chen et al. 2013); the Krüppel-like factor 6 (KLF6; Ueharu et al. 2014); and the juxtacrine signaling molecule ephrin-B2 (Yoshida et al. 2015). A number of the pituitary stem cell markers identified may be involved in constructing and maintaining the 3D network, in particular the cadherins, ephrins and CAR.

An intriguing question is how the multiple markers are distributed within the stem cell population. Regarding the embryonic markers, it can be assumed that different stages along the stem cell differentiation path are typified by different factors in a sequence similar to the embryogenic process. Accordingly, it is proposed that the pituitary stem cell compartment represents a heterogeneous pool of cells in different stages of life cycle and activation (Fig. 1), with the different markers, alone or in combination, designating the different phases, varying from quiescent stem cell states to committed precursor steps (reviewed in Vankelecom 2010, and Vankelecom and Chen 2014). Such heterogeneity within the stem cell population has also increasingly been demonstrated in other organs (Bond et al. 2015; Donati and Watt 2015). Single-cell transcriptomic analysis of brain stem cells identified the presence of four subgroups in different stages of activation (most prominently in response to injury), ranging from dormant to primed-quiescent, active and active-dividing stem cells (Llorens-Bobadilla et al. 2015). Of note, pituitary and brain stem cells show striking similarities, in particular regarding their localization around a lumen (cleft and ventricle, respectively) and their expression of common markers (such as SOX2, NESTIN and S100; Kriegstein and Alvarez-Buylla 2009). In addition, different stem cell subpopulations may be distributed over the different proposed niches in the pituitary. For instance, stem cells in the parenchymal SOX2$^+$ cell clusters may be further advanced in their 'priming' (as compared to the MZ cells) to allow swifter contribution to the neigboring parenchymal endocrine tissue, or they may only form specific cell types more prevalent at the particular parenchymal spots.

Apart from their labeling value, the factors mentioned must, of course, also have a function in the pituitary stem cells. Not much is known yet about these duties, but ideas may be inferred from other stem cells or from pituitary embryogenesis. The transcription factors SOX2, OCT4, NANOG and KLF4 most likely play a role similar to the one in ES and iPS cells (De Los Angeles et al. 2015; Hyslop et al. 2005; Takahashi and Yamanaka 2006), i.e., to keep the pituitary stem cells in an undifferentiated, multipotent state. Genetic disruptions of SOX2 leads to certain forms of hormonal cell deficiency and pituitary hypoplasia (Jayakody et al. 2012; Kelberman et al. 2008), at least partly due to a reduced expansion and function of the RP progenitor cells, thereby giving some hint toward the importance of SOX2 in adult stem cell regulation. NOTCH signaling is likely also involved in the control of stem cell proliferation in the adult gland, since NOTCH downregulation in the embryonic pituitary coincides with cell cycle exit of the RP progenitor cells and genetic ablation of NOTCH signaling in the developing pituitary results in severe AP hypoplasia due to reduced proliferative capacity of RP progenitors (Kita et al. 2007; Monahan et al. 2009; Raetzman et al. 2004; Tando et al. 2013; Zhu et al. 2005, 2007). In addition, stem cells increase in number in AP cell cultures upon NOTCH activation (Chen et al. 2006; Tando et al. 2013). SOX2 and NOTCH (together with PROP1) may form an interacting signaling network within the stem cell compartment (reviewed in Willems and Vankelecom 2014).

In the adult pituitary, PROP1 may be required in the progressing stem/progenitor cells to enable migration from the MZ or the in-gland stem cell clusters toward the parenchyma for further differentiation. This proposed role is based on knowledge from embryogenesis, where genetic inactivation of PROP1 results in failure of RP progenitors to migrate from the proliferative progenitor region to the developing AP, resulting in extensive pituitary hypoplasia and absence of nearly all AP cell lineages (Himes and Raetzman 2009; Ward et al. 2005; Zhu et al. 2005, 2007). The prominent presence of the epithelial marker CDH1 and the tight-junction protein CAR suggests a role for these factors in stitching the marginal and parenchymal niches of stem cells together in a 3D network (Fig. 1). Escape from this 'imprisonment' would require conversion of the organized connected epithelial cell type to the untied and motile mesenchymal cell phenotype through the process of epithelial-mesenchymal transition (EMT; Kalluri and Weinberg 2009; Vankelecom and Chen 2014). PROP1 may be required for EMT since it downregulates CDH1 by activating the expression of the CDH1 repressor SNAI2/SLUG in the RP progenitor zone (Himes and Raetzman 2009). Stem cell migration in the adult gland may further be regulated by CXCR4, which is known for its chemotactic activity and regulation of EMT (Hu et al. 2014; Kalluri and Weinberg 2009). Also NESTIN[+] pituitary cells have previously been shown to possess motile capacity (in vitro) as a possible result of EMT (Krylyshkina et al. 2005). Finally, ephrin signaling may play a role in the formation and organization of the proposed 3D stem cell network through actions of attraction and repulsion (Solanas and Batlle 2011). Taken together, a number of the factors identified seem to be involved in regulating the balance between movement and bonding of the pituitary stem cells.

In conclusion, the pituitary stem cell phenotyping list is progressively expanding. Although one can speculate on the role these markers play in adult pituitary stem cell regulation, a firm foundation remains to be provided.

Pituitary Stem Cells: Uncovering the Hidden Functions

As mentioned above, new tissue cells can be generated by the resident stem cells during homeostatic turnover, adaptive responses to body requirements and regenerative reactions to tissue damage (Alvarez-Buylla and Lim 2004; Rando 2006; Slack 2008; Vankelecom 2012; Vankelecom and Chen 2014; Wabik and Jones 2015). Whether stem cells in the pituitary also form or renew hormonal cells during postnatal life is a tempting question. Given the slow turnover rate of the adult gland (Levy 2002, 2008; Nolan et al. 1998; Rando 2006; Vankelecom 2012; Vankelecom and Chen 2014), the involvement of stem cells would be most evident in situations of more activated cell remodeling in the tissue. Recently, studies have started to explore the functional position of pituitary stem cells in more dynamic conditions, including neonatal maturation, adaptation to endocrine demands, regeneration after damage, and tumor formation.

Neonatal Pituitary Maturation

The (rodent) pituitary undergoes a substantial growth phase immediately after birth, with increasing numbers of hormonal cells developing during the first neonatal week(s) (Melmed 2010; Vankelecom 2010, 2012; Vankelecom and Chen 2014). During this period, the pituitary stem cell population resides in a state of activation as compared to later in life, showing higher cell numbers, higher proliferation rate and higher stem cell functionality (i.e., sphere-forming and multipotent differentiation capacity; Chen et al. 2005, 2009; Gremeaux et al. 2012). Moreover, stemness and embryonic markers are prominently expressed in neonatal stem cells (Chen et al. 2009; Gremeaux et al. 2012; Kikuchi et al. 2007). The topography of the $SOX2^+$ cells further underlines this higher activation status, with increased numbers of clusters in the neonatal AP lobe and higher abundance of $SOX2^+$ cells at the junctions of the AP and IL (wedges), with signs of $SOX2^+$ cells sprouting from these putative germinal regions (Gremeaux et al. 2012; Fig. 1). Neonatal $SOX2^+$ cell clusters in the vicinity of the MZ are clearly connected to the MZ (Gremeaux et al. 2012), supporting the idea that the cell clusters observed in the AP lobe originate from this zone (Fig. 1). Of note, SOX2 is mainly not observed together with hormones, which may be interpreted again as mutual exclusion, i.e., with differentiation only occurring when the guardians of multipotency like SOX2 are downregulated or extinguished.

Tracing of SOX2$^+$ cells from embryonic to neonatal age (technically speaking, by tamoxifen induction in pregnant Sox2^{CreERT2}/R26YFP$^{flox/flox}$ mice) revealed traced 'yellow fluorescent protein'-positive (YFP$^+$) cells in the neonatal pituitary that express early and late differentiation markers of hormonal cells, indicating that embryonic SOX2-expressing cells contribute to neonatal hormonal cells (Andoniadou et al. 2013; Rizzoti et al. 2013). It should be realized that this finding does not automatically mean that the neonatal stem cells per se drive the early expansion phase (by forming new hormonal cells) during those first weeks after birth. Short-term SOX2$^+$ lineage tracing should, therefore, be done within the neonatal period itself. YFP was also found together with SOX2, indicating that the embryonic SOX2$^+$ progenitor cells persist after birth (e.g., by themselves or after self-renewal). Also in other tissues, SOX2$^+$ postnatal (stem) cells originate from fetal SOX2$^+$ (progenitor) cells (Arnold et al. 2011). This idea is also in line with the embryonic phenotype of the postnatal pituitary stem cells, as supported by the marker expression profile (see above).

Taken together, the activated nature of the stem cell compartment during the first weeks after birth suggests a dynamic participation in the neonatal maturation process of the gland, although most evidence so far is circumstantial.

Basal and Adapting Adult Pituitary

Basal Turnover The contribution of stem cells to the slow homeostatic turnover in the postnatal gland was only recently demonstrated (Andoniadou et al. 2013; Rizzoti et al. 2013). Tamoxifen-induced SOX2$^+$ or SOX9$^+$ lineage tracing starting from four to eight weeks of age revealed the existence of hormone$^+$ cells derived from the traced (YFP$^+$) cells as analyzed 8 to 14 months later. However, their number was small, which in the first place reflects the low turnover rate of the adult gland under physiological conditions but at the same time suggests that the contribution of stem cells to new hormonal cells is only very limited under basal conditions. Endocrine cell turnover in the basal pituitary would thus mainly rely on proliferation of differentiated hormonal cells (Langlais et al. 2013; Vankelecom and Chen 2014). The vast majority of the YFP$^+$ cells were still SOX2$^+$/SOX9$^+$ after the long-term tracing, supporting a long-lived stem cell phenotype of high quiescence and/or a persistent (but slow) self-renewal activity. Taken together, the findings from lineage tracing do not support a major input of stem cells in adult pituitary homeostatic turnover, which is in line with findings in other 'lazy' tissues (Barker et al. 2010; Huch et al. 2013a,b; Rando 2006; Slack 2008; Vankelecom 2012; Wabik and Jones 2015).

Plastic Cell Adaptations In contrast to the quite immeasurable cell neogenesis under basal conditions, the pituitary's cell composition more actively changes in response to peripheral signals conveying endocrine needs. As a prominent example, the number (and activity) of lactotropes rises during pregnancy and lactation to meet the heightened demand for PRL (Haggi et al. 1986; Vankelecom 2012). This expansion is at least partly due to elevated estrogen levels in these conditions. The involvement of stem cells may be concluded from the recent observation that short-term estradiol treatment (of male mice) causes a 10-fold increase in dividing $SOX2^+$ cells. The total number of $SOX2^+$ cells, however, did not change, suggesting that the generated daughter cells immediately stopped expressing SOX2 to differentiate into the demanded lactotropes (Rizzoti et al. 2013). To underpin a direct contribution of stem cells to the expanding lactotrope population, lineage-tracing experiments are further needed. Whether the stem cells also play a role in the rise of somatotropes during puberty, or in the (continuous) adaptation of gonadotropes during sexual maturation and estrous cycling, is at present unknown.

Other forms of enhanced pituitary cell remodeling are seen when negative feedback dissipates because of ablation of target organs. Adrenalectomy causes a swift, transient rise in corticotropes, whereas gonadectomy triggers a fast and transitory increase in gonadotropes. Previous studies provided circumstantial evidence that 'hormonally null cells' (the at that time postulated pituitary stem cells) contribute to the new corticotropes and gonadotropes (Levy 2002, 2008; Nolan et al. 1998; Nolan and Levy 2006). Recent studies provided more direct evidence that pituitary stem cells are involved. The $SOX2^+$ cells along the cleft expand after adrenalectomy (Langlais et al. 2013). In addition, $SOX9^+$ lineage tracing showed that ~20 % of the newborn corticotropes are derived from the $SOX9^+$ stem cells (Rizzoti et al. 2013). The other new $ACTH^+$ cells may be produced by corticotrope proliferation (see also Langlais et al. 2013). After gonadectomy, dividing $SOX2^+$ cells increase by four-fold but total $SOX2^+$ cell numbers do not change (Rizzoti et al. 2013), again suggesting that the generated progenitor cells immediately differentiate with prompt disappearance of SOX2.

Taken together, recent studies provide supportive evidence that adult pituitary stem cells have the capacity to differentiate into hormonal cells in vivo under challenging physio- and pathological conditions.

Pituitary Regeneration and Impact of Aging

A number of adult tissues are capable of restoring cells following destruction by physical or chemical impacts (Rando 2006; Vankelecom 2012; Vankelecom and Chen 2014; Willems and Vankelecom 2014; Wabik and Jones 2015). In several of these regenerative responses (like in muscle), stem cells are mobilized and directly involved to generate the new cells (Conboy and Rando 2005). Also in slow-turnover organs like liver and pancreas, hidden ('facultative') stem cells are activated under certain damaging conditions to drive the regenerative reaction (Barker et al. 2010; Huch et al. 2013a,b; Rando 2006; Slack 2008; Vankelecom 2012; Wabik and Jones 2015; Xu et al. 2008).

Regarding the pituitary, it has only recently been established that the adult, mature gland has the potential to regenerate cells after destruction. Through a transgenic mouse approach, damage was inflicted in the pituitary by killing the somatotrope (GH^+) cells using diphtheria toxin (DT; Fu et al. 2012; Luque et al. 2011). The injury triggered an immediate response of the pituitary stem cells, which started to expand in number and to co-express GH (Fu et al. 2012). Five to six months after ablation, the somatotrope cell number was significantly restored (up to 60 %). The study for the first time showed that the pituitary, when suffering damage at adult age, has the potential to regenerate destroyed tissue (Fu et al. 2012). In addition, it advanced the stem cells as the likely drivers of regeneration (Fig. 1) and source of the newborn somatotropes. Meanwhile, this regenerative capability appeared more general and not limited to somatotropes. In an analogous model in which lactotropes were destroyed with DT (Fu and Vankelecom 2012), restoration was also observed (up to ~60 %) although clearly faster (already after four to six weeks), which may be due to the involvement and cooperation of more than one process. Stem cell participation with expansion and PRL co-expression was again observed, but in addition there was enhanced proliferation of the surviving (or newly formed) PRL^+ cells as well as increased numbers of double PRL^+/GH^+ cells, suggestive of an activated transdifferentiaton process of somatotropes toward lactotropes (Fu and Vankelecom 2012). Finally, pituitary stem cells also expanded in number following gonadotrope ablation by DT, but detailed analysis of this mouse model was not possible because of cardiotoxicity (Vankelecom, unpublished observations).

A recent follow-up characterization study of the somatotrope ablation-and-regeneration model revealed some interesting features of the pituitary's regenerative capacity (Willems et al. 2016). First, the regeneration level appeared to be capped (at ~60 %, or in other words, restoration to ~70 % of the normal GH^+ cell number), even if the recovery period was largely extended from 0.5 to 1.5 year. Either the regenerative power of the pituitary does not go beyond certain levels or higher restoration is not needed to reach sufficient 'physiological' GH activity. Serum GH concentrations were restored to about one-third of normal values, at the same time indicating that regeneration does not only occur at the morphological level but also, although in a more limited fashion, at the functional (hormone-secretory) level. Furthermore, and rather surprisingly, the restorative capacity of the pituitary fades very fast at aging; middle-aged mice (eight months old) no longer showed recovery (as compared to eight-week-old mice), not even after long recuperation periods (Willems et al. 2016). Interestingly, this disappearance of regenerative competence coincides with a decline in pituitary stem cell number. Moreover, the stem cells of the older pituitary are less talented in generating spheres (as a functional characteristic of stem cells), which decrease in number and size, although differentiation to hormonal cell lineages still occurs in the spheres. $SOX2^+$ signals in the spheres from the older pituitaries were most of the time found in the cytoplasm and not in the nucleus of the cells, where SOX2 should be present to maintain the stem cell phenotype (Willems et al. 2016). Taken together, these findings suggest a decrease in overall fitness of the pituitary stem cells at aging - not being maintained in a primitive state - which may lead to fast (-er) exhaustion of their potential during differentiation and regenerative attempts. In other tissues (e.g., muscle and heart), the stem cell population is also negatively affected by age, undergoing a decline in number and regenerative capacity (Blau et al. 2015; Hariharan and Sussman 2015).

Further intriguingly, restoration was not observed anymore when the injury impact was prolonged by extending the DT injection period (from 3 to 10 days), although the somatotrope ablation grade obtained was identical (Willems et al. 2016). The stem cell compartment still reacted to the prolonged injury by promptly expanding (although somewhat less than after the short-term injury impact), but no co-expression of GH was found despite the fact that the stem cells were still capable of differentiating into all hormonal cell types when assessed in vitro (using pituispheres; Willems et al. 2016). As found in other tissues (like the hippocampus and the hematopoietic system), the regenerative power may become exhausted when subsequent attempts are over and again suffocated during the long-term impact (Botnick et al. 1979; Sierra et al. 2015). Stem cells may have reached their expansive limit or crossed their threshold of restorative competence, or the reacting, regenerating stem cell pool may become depleted while possibly other, more quiet, stem cell populations remain unaffected (which would explain the preservation of sphere-forming and multipotent capacity). Alternatively, the observation of a remaining intrinsic functionality after prolonged DT treatment may suggest that a deficiency in stem cell regulatory networks, as emanating from the niche, pituitary parenchyma or systemic circulation, may lie at the basis of the regenerative failure.

Finally, our recent characterization study started to search for molecular mechanisms underlying regeneration and exposed some embryonic, stemness and repairing pathways that may be involved in the stem cell reaction to injury, in particular EMT, growth factor (FGF and epidermal growth factor, EGF) and Hippo pathways (Willems et al. 2016). Activation of FGF and EGF can lead to increased pituitary stem cell numbers, as has been demonstrated in AP cell aggregate cultures (Chen et al. 2006).

Taken together, recent studies support a function for pituitary stem cells in pituitary regeneration (Fig. 1). Activation of the stem cells and movement to the site of injury may be steered by embryonic, proliferative (growth factor), migratory (EMT) and restorative (Hippo) signaling pathways. This regenerative capacity appears not boundless but limited both in age-related terms and final efficacy.

Pituitary Tumorigenesis

As mentioned, the process of tumorigenesis in the pituitary remains far from understood (Melmed, 2003, 2011; Vankelecom and Gremeaux 2010; Vankelecom 2012). An appealing but largely untouched question is about the position of the pituitary stem cells during the tumorigenic event in the tissue. Furthermore, are

TSC present in pituitary tumors, and if so, are they linked to the resident stem cells? Previous studies regarding this subject have been extensively reviewed before (Florio 2011; Lloyd et al. 2013; Vankelecom and Gremeaux 2010; Vankelecom 2012; Vankelecom and Chen 2014). Some candidate TSC were proposed, for instance, based on spheroid formation (Xu et al. 2009) and marker expression (Barbieri et al. 2008), but convincing evidence was not provided and some results remained questionable (discussed in Vankelecom and Gremeaux 2010; Vankelecom 2012; Vankelecom and Chen 2014).

A recent study of our group found that pituitary adenomas (as obtained from human patients) contain a SP with prominent expression of tumor stemness markers (like CXCR4) and stemness signaling pathways (like EMT) and that enriches for cells that form tumorspheres in a self-renewing sequence (Mertens et al. 2015). The pituitary tumor SP (as analyzed for the AtT20 cell line) showed tumor-growth advantage in vivo in immunodeficient mice. Thus, the pituitary tumor SP holds molecular and functional characteristics supporting a TSC phenotype. In addition, CXCR4 signaling may be involved in AtT20 tumorigenesis since inhibition of the pathway reduced tumor size in vivo as well as EMT activity (cell motility) in vitro (Mertens et al. 2015). Moreover, the AtT20 SP showed resistance to the chemo-therapeutic drug temozolomide (Mertens and Vankelecom, unpublished observations), further supporting a TSC phenotype. Also interestingly, the SP of pituitary tumors displays some appealing molecular differences with the candidate TSC (SP) of malignant cancer types (melanoma and pancreatic cancer; Van den Broeck et al. 2013; Wouters et al. 2013) such as an upregulated senescence program, which might explain why pituitary tumors typically remain benign (Mertens and Vankelecom, unpublished observations).

The pituitary tumor SP also shows upregulated expression of *SOX2* (Mertens et al. 2015), which may indicate either a link between the stem cells and the putative TSC or simply activation of SOX2 expression in the candidate TSC. Moreover, pituitaries from a mouse model in which pituitary (PRL^+) tumors develop in situ (i.e., the dopamine receptor D2 knockout or $Drd2^{-/-}$ mouse) contains more SP and $SOX2^+$ cells than wildtype glands (Mertens et al. 2015). This observation is in accordance with the presence of a TSC (characterized by SP and $SOX2^+$ phenotype) in the mouse pituitary tumors, adding up to the SP and $SOX2^+$ cells of the surrounding normal tissue. In addition, or alternatively, the observation may point to an activated and expanded 'normal' stem cell compartment when tumorigenesis is occurring in the gland. Regarding the latter idea, it is not known yet what the consequence of pituitary stem cell activation may be. As already mentioned above, it is assumed that disappearance of SOX2 from the nucleus (by exclusion or active expulsion) is needed to allow differentiation (Chen et al. 2009; Fu et al. 2012; Vankelecom and Chen 2014; Willems and Vankelecom 2014), as has been demonstrated in ES cells (Baltus et al. 2009). The observation in $Drd2^{-/-}$ pituitaries of a predominant increase in $SOX2^+$ cells in which SOX2 is present in the cytoplasm (Mertens et al. 2015) might support a high(-er) differentiation rate toward the tumor PRL^+ cells.

Alternatively, stem cell activation may represent a defence reaction or may paradoxically activate the TSC or feed the tumor by paracrine influences. An indirect role of SOX2$^+$ stem cells as paracrine tumor-activating cells has also been suggested in a mouse model of adamantinomatous craniopharyngioma (Andoniadou et al. 2013; Gaston-Massuet et al. 2011; see also Andoniadou 2016), a pituitary tumor that originates from ectopic remnants of RP and in that way clearly differs from the typical AP tumors. On the other hand, resident stem cells may directly generate the TSC, as has been shown in some other tumor types (e.g., of intestine, skin and brain; Barker et al. 2009; Chen et al. 2012; Lapouge et al. 2011; Schepers et al. 2012). Stem cell lineage tracing in mouse models developing typical pituitary tumors is now needed to explore the link between tumorigenesis, TSC and stem cells in the pituitary.

Taken together, another hidden function of the pituitary stem cells may reside in the process of tumorigenesis (Fig. 1), either as the direct creators of the TSC or as the reacting compartment activated in response to the tumorigenic assault occurring in the tissue, resulting in a paracrine impact that likely is intended to defend the tissue but may eventually fuel the tumor.

Conclusion and General Perspectives

Recent studies have started to unveil the hidden functions of pituitary stem cells. Their job appears to primarily emerge in the active or challenged pituitary, i.e., during neonatal maturation, tumorigenesis and repair of damage (Fig. 1). The stem cells seem much less involved in the more subtle adaptations during basic turnover. Aging has an early and negative impact on the number and fitness of the pituitary stem cells, likely explaining the regenerative failure with advancing age. From the currently available data, embryonic programs appear to be recycled for postnatal stem cell regulation, activation and differentiation. Further efforts are needed to pinpoint the molecular mechanisms underlying the stem cell functions. An additional intriguing aspect to be deciphered is how the newborn endocrine cells topographically and functionally integrate into the (existing) hormonal cell networks.

More insight into the pituitary stem cell role and regulation may in the end advance the treatment of hypopituitarism patients, particularly within the context of regenerative medicine (extensively reviewed in Vankelecom and Chen 2014 and Willems and Vankelecom 2014). Pituitary stem cells may provide life-long cures for mutation-, surgery- and/or trauma-induced pituitary deficiencies. The endogenous stem cells may be stimulated to restore or repair the defective tissue, or the missing hormonal cells may be generated from stem cells ex vivo and implanted. The approaches are expected to be superior to HRT, which generates artificial hormone levels and cannot mimic hormone secretory cyclicity and pulsatility. However, many pressing issues remain, particularly regarding functional integration and safety, before moving to translational applications to cure conditions of hypopituitarism.

In sum, pituitary stem cells represent potential protagonists in the gland with still mysterious functions in pituitary biology and pathology, although recent studies have started to 'write' some possible scripts. Clearly, there remains a long way to go in the still under-explored domain of pituitary stem cell functions and associated clinical opportunities.

References

Allaerts W, Vankelecom H (2005) History and perspectives of pituitary folliculo-stellate cell research. Eur J Endocrinol 153:1–12

Alvarez-Buylla A, Lim DA (2004) For the long run: maintaining germinal niches in the adult brain. Neuron 41:683–686

Andoniadou CL (2016) Pituitary stem cells during normal physiology and disease. In: Pfaff D, Christen Y (eds) Stem cells in neuroendocrinology. Springer, Heidelberg

Andoniadou CL, Matsushima D, Mousavy Gharavy SN, Signore M, Mackintosh AI, Schaeffer M, Gaston-Massuet C, Mollard P, Jacques TS, Le Tissier P, Dattani MT, Pevny LH, Martinez-Barbera JP (2013) Sox2(+) stem/progenitor cells in the adult mouse pituitary support organ homeostasis and have tumor-inducing potential. Cell Stem Cell 13:433–445

Arnold K, Sarkar A, Yram MA, Polo JM, Bronson R, Sengupta S, Seandel M, Geijsen N, Hochedlinger K (2011) Sox2(+) adult stem and progenitor cells are important for tissue regeneration and survival of mice. Cell Stem Cell 9:317–329

Baltus GA, Kowalski MP, Zhai H, Tutter AV, Quinn D, Wall D, Kadam S (2009) Acetylation of sox2 induces its nuclear export in embryonic stem cells. Stem Cells 27:2175–2184

Barbieri F, Bajetto A, Stumm R, Pattarozzi A, Porcile C, Zona G, Dorcaratto A, Ravetti JL, Minuto F, Spaziante R, Schettini G, Ferone D, Florio T (2008) Overexpression of stromal cell-derived factor 1 and its receptor CXCR4 induces autocrine/paracrine cell proliferation in human pituitary adenomas. Clin Cancer Res 14:5022–5032

Barker N, Ridgway R, van Es JH, van de Wetering M, Begthel H, van den Born M, Danenberg E, Clarke AR, Sansom OJ, Clevers H (2009) Crypt stem cells as the cells-of-origin of intestinal cancer. Nature 457:608–611

Barker N, Bartfeld S, Clevers H (2010) Tissue-resident adult stem cell populations of rapidly self-renewing organs. Cell Stem Cell 7:656–670

Bilodeau S, Roussel-Gervais A, Drouin J (2009) Distinct developmental roles of cell cycle inhibitors p57Kip2 and p27Kip1 distinguish pituitary progenitor cell cycle exit from cell cycle reentry of differentiated cells. Mol Cell Biol 29:1895–1908

Blau HM, Cosgrove BD, Ho AT (2015) The central role of muscle stem cells in regenerative failure with aging. Nat Med 21:854–862

Bond AM, Ming GL, Song H (2015) Adult mammalian neural stem cells and neurogenesis: five decades later. Cell Stem Cell 17:385–395

Botnick LE, Hannon EC, Hellman S (1979) Nature of the hemopoietic stem cell compartment and its proliferative potential. Blood Cells 5:195–210

Chauvet N, El-Yandouzi T, Mathieu MN, Schlernitzauer A, Galibert E, Lafont C, Le Tissier P, Robinson IC, Mollard P, Coutry N (2009) Characterization of adherens junction protein expression and localization in pituitary cell networks. J Endocrinol 202:375–387

Chen J, Hersmus N, Van Duppen V, Caesens P, Denef C, Vankelecom H (2005) The adult pituitary contains a cell population displaying stem/progenitor cell and early embryonic characteristics. Endocrinology 146:3985–3998

Chen J, Crabbe A, Van Duppen V, Vankelecom H (2006) The notch signaling system is present in the postnatal pituitary: marked expression and regulatory activity in the newly discovered side population. Mol Endocrinol 20:3293–3307

Chen J, Gremeaux L, Fu Q, Liekens D, Van Laere S, Vankelecom H (2009) Pituitary progenitor cells tracked down by side population dissection. Stem Cells 27:1182–1195

Chen J, Li Y, Yu TS, McKay RM, Burns DK, Kernie SG, Parada LF (2012) A restricted cell population propagates glioblastoma growth after chemotherapy. Nature 488:522–526

Chen M, Kato T, Higuchi M, Yoshida S, Yako H, Kanno N, Kato Y (2013) Coxsackievirus and adenovirus receptor-positive cells compose the putative stem/progenitor cell niches in the marginal cell layer and parenchyma of the rat anterior pituitary. Cell Tissue Res 354:823–836

Clevers H (2011) The cancer stem cell: premises, promises and challenges. Nat Med 17:313–319

Conboy IM, Rando TA (2005) Aging, stem cells and tissue regeneration - Lessons from muscle. Cell Cycle 4:407–410

Dalerba P, Cho RW, Clarke MF (2007) Cancer stem cells: models and concepts. Annu Rev Med 58:267–284

De Los AA, Ferrari F, Xi R, Fujiwara Y, Benvenisty N, Deng H, Hochedlinger K, Jaenisch R, Lee S, Leitch HG, Lensch MW, Lujan E, Pei D, Rossant J, Wernig M, Park PJ, Daley GQ (2015) Hallmarks of pluripotency. Nature 525:469–478

Donati G, Watt FM (2015) Stem cell heterogeneity and plasticity in epithelia. Cell Stem Cell 16:465–476

Ericson J, Norlin S, Jessell TM, Edlund T (1998) Integrated FGF and BMP signaling controls the progression of progenitor cell differentiation and the emergence of pattern in the embryonic anterior pituitary. Development 125:1005–1015

Fauquier T, Guerineau NC, McKinney RA, Bauer K, Mollard P (2001) Folliculostellate cell network: a route for long-distance communication in the anterior pituitary. Proc Natl Acad Sci USA 98:8891–8896

Fauquier T, Rizzoti K, Dattani M, Lovell-Badge R, Robinson IC (2008) SOX2-expressing progenitor cells generate all of the major cell types in the adult mouse pituitary gland. Proc Natl Acad Sci USA 105:2907–2912

Florio T (2011) Adult pituitary stem cells: from pituitary plasticity to adenoma development. Neuroendocrinology 94:265–277

Fu Q, Vankelecom H (2012) Regenerative capacity of the adult pituitary: multiple mechanisms of lactotrope restoration after transgenic ablation. Stem Cells Dev 21:3245–3257

Fu Q, Gremeaux L, Luque RM, Liekens D, Chen J, Buch T, Waisman A, Kineman R, Vankelecom H (2012) The adult pituitary shows stem/progenitor cell activation in response to injury and is capable of regeneration. Endocrinology 153:3224–3235

Fujiwara K, Kikuchi M, Takigami S, Kouki T, Yashiro T (2007) Expression of retinaldehyde dehydrogenase 1 in the anterior pituitary glands of adult rats. Cell Tissue Res 329:321–327

Garcia-Lavandeira M, Quereda V, Flores I, Saez C, Diaz-Rodriguez E, Japon MA, Ryan AK, Blasco MA, Dieguez C, Malumbres M, Alvarez CV (2009) A GRFa2/Prop1/stem (GPS) cell niche in the pituitary. PLoS One 4, e4815

Gasco V, Prodam F, Pagano L, Grottoli S, Belcastro S, Marzullo P, Beccuti G, Ghigo E, Aimaretti G (2012) Hypopituitarism following brain injury: when does it occur and how best to test? Pituitary 15:20–24

Gasco V, Prodam F, Grottoli S, Marzullo P, Longobardi S, Ghigo E, Aimaretti G (2013) GH therapy in adult GH deficiency: a review of treatment schedules and the evidence for low starting doses. Eur J Endocrinol 168:R55–66

Gaston-Massuet C, Andoniadou CL, Signore M, Jayakody SA, Charolidi N, Kyeyune R, Vernay B, Jacques TS, Taketo MM, Le Tissier P, Dattani MT, Martinez-Barbera JP (2011) Increased Wingless (Wnt) signaling in pituitary progenitor/stem cells gives rise to pituitary tumors in mice and humans. Proc Natl Acad Sci USA 108:11482–11487

Gremeaux L, Fu Q, Chen J, Vankelecom H (2012) Activated phenotype of the pituitary stem/ progenitor cell compartment during the early-postnatal maturation phase of the gland. Stem Cells Dev 21:801–813

Haggi ES, Al T, Maldonado CA, Aoki A (1986) Regression of redundant lactotrophs in rat pituitary gland after cessation of lactation. J Endocrinol 111:367–373

Hariharan N, Sussman MA (2015) Cardiac aging - getting to the stem of the problem. J Mol Cell Cardiol 83:32–36

Himes AD, Raetzman LT (2009) Premature differentiation and aberrant movement of pituitary cells lacking both Hes1 and Prop1. Dev Biol 325:151–161

Horiguchi K, Ilmiawati C, Fujiwara K, Tsukada T, Kikuchi M, Yashiro T (2012) Expression of chemokine CXCL12 and its receptor CXCR4 in folliculostellate (FS) cells of the rat anterior pituitary gland: the CXCL12/CXCR4 axis induces interconnection of FS cells. Endocrinology 153:1717–1724

Hu TH, Yao Y, Yu S, Han LL, Wang WJ, Guo H, Tian T, Ruan ZP, Kang XM, Wang J, Wang SH, Nan KJ (2014) SDF-1/CXCR4 promotes epithelial-mesenchymal transition and progression of colorectal cancer by activation of the Wnt/β-catenin signaling pathway. Cancer Lett 354:417–426

Huch M, Boj SF, Clevers H (2013a) Lgr5(+) liver stem cells, hepatic organoids and regenerative medicine. Regen Med 8:385–387

Huch M, Bonfanti P, Boj SF, Sato T, Loomans CJ, van de Wetering M, Sojoodi M, Li VS, Schuijers J, Gracanin A, Ringnalda F, Begthel H, Hamer K, Mulder J, van Es JH, de Koning E, Vries RG, Heimberg H, Clevers H (2013b) Unlimited in vitro expansion of adult bi-potent pancreas progenitors through the Lgr5/R-spondin axis. EMBO J 32:2708–2721

Hyslop LA, Armstrong L, Stojkovic M, Lako M (2005) Human embryonic stem cells: biology and clinical implications. Expert Rev Mol Med 7:1–21

Jayakody SA, Andoniadou CL, Gaston-Massuet C, Signore M, Cariboni A, Bouloux PM, Le Tissier P, Pevny LH, Dattani MT, Martinez-Barbera JP (2012) SOX2 regulates the hypothalamic-pituitary axis at multiple levels. J Clin Invest 122:3635–3646

Jensen JN, Cameron E, Garay MVR, Starkey TW, Gianani R, Jensen J (2005) Recapitulation of elements of embryonic development in adult mouse pancreatic regeneration. Gastroenterology 128:728–741

Kalluri R, Weinberg RA (2009) The basics of epithelial-mesenchymal transition. J Clin Invest 119:1420–1428

Kelberman D, de Castro SC, Huang S, Crolla JA, Palmer R, Gregory JW, Taylor D, Cavallo L, Faienza MF, Fischetto R, Achermann JC, Martinez-Barbera JP, Rizzoti K, Lovell-Badge R, Robinson IC, Gerrelli D, Dattani MT (2008) SOX2 plays a critical role in the pituitary, forebrain, and eye during human embryonic development. J Clin Endocrinol Metab 93:1865–1873

Kelberman D, Rizzoti K, Lovell-Badge R, Robinson IC, Dattani MT (2009) Genetic regulation of pituitary gland development in human and mouse. Endocr Rev 30:790–829

Kikuchi M, Yatabe M, Kouki T, Fujiwara K, Takigami S, Sakamoto A, Yashiro T (2007) Changes in E- and N-cadherin expression in developing rat adenohypophysis. Anat Rec (Hoboken) 290:486–490

Kita A, Imayoshi I, Hojo M, Kitagawa M, Kokubu H, Ohsawa R, Ohtsuka T, Kageyama R, Hashimoto N (2007) Hes1 and Hes5 control the progenitor pool, intermediate lobe specification, and posterior lobe formation in the pituitary development. Mol Endocrinol 21:1458–1466

Kriegstein A, Alvarez-Buylla A (2009) The glial nature of embryonic and adult neural stem cells. Annu Rev Neurosci 32:149–184

Krylyshkina O, Chen J, Mebis L, Denef C, Vankelecom H (2005) Nestin-immunoreactive cells in rat pituitary are neither hormonal nor typical folliculo-stellate cells. Endocrinology 146:2376–2387

Langlais D, Couture C, Kmita M, Drouin J (2013) Adult pituitary cell maintenance: lineage specific contribution of self-duplication. Mol Endocrinol 27:1103–1112

Lapouge G, Youssef KK, Vokaer B, Achouri Y, Michaux C, Sotiropoulou PA, Blanpain C (2011) Identifying the cellular origin of squamous skin tumors. Proc Natl Acad Sci USA 108:7431–7436

Le Tissier PR, Hodson DJ, Lafont C, Fontanaud P, Schaeffer M, Mollard P (2012) Anterior pituitary cell networks. Front Neuroendocrinol 33:252–266

Levy A (2002) Physiological implications of pituitary trophic activity. J Endocrinol 174:147–155

Levy A (2008) Stem cells, hormones and pituitary adenomas. J Neuroendocrinol 20:139–140

Llorens-Bobadilla E, Zhao S, Baser A, Saiz-Castro G, Zwadlo K, Martin-Villalba A (2015) Single-cell transcriptomics reveals a population of dormant neural stem cells that become activated upon brain injury. Cell Stem Cell 17:329–340

Lloyd RV, Hardin H, Montemayor-Garcia C, Rotondo F, Syro LV, Horvath E, Kovacs K (2013) Stem cells and cancer stem-like cells in endocrine tissues. Endocr Pathol 24:1–10

Luque RM, Lin Q, Cordoba-Chacon J, Subbaiah PV, Buch T, Waisman A, Vankelecom H, Kineman RD (2011) Metabolic impact of adult-onset, isolated, growth hormone deficiency (AOiGHD) due to destruction of pituitary somatotropes. PLoS One 6, e15767

Mehta A, Dattani MT (2008) Developmental disorders of the hypothalamus and pituitary gland associated with congenital hypopituitarism. Best Pract Res Clin Endocrinol Metab 22:191–206

Melmed S (2003) Mechanisms for pituitary tumorigenesis: the plastic pituitary. J Clin Invest 112:1603–1618

Melmed S (2010) The pituitary, 3rd edn. Academic, New York

Melmed S (2011) Pathogenesis of pituitary tumors. Nat Rev Endocrinol 7:257–266

Mertens F, Gremeaux L, Chen J, Fu Q, Willems C, Roose H, Govaere O, Roskams T, Cristina C, Becú-Villalobos D, Jorissen M, Poorten VV, Bex M, van Loon J, Vankelecom H (2015) Pituitary tumors contain a side population with tumor stem cell-associated characteristics. Endocr Relat Cancer 22:481–504

Mollard P, Hodson DJ, Lafont C, Rizzoti K, Drouin J (2012) A tridimensional view of pituitary development and function. Trends Endocrinol Metab 23:261–269

Monahan P, Rybak S, Raetzman LT (2009) The notch target gene HES1 regulates cell cycle inhibitor expression in the developing pituitary. Endocrinology 150:4386–4394

Nolan LA, Levy A (2006) A population of non-luteinising hormone/non-adrenocorticotrophic hormone-positive cells in the male rat anterior pituitary responds mitotically to both gonadectomy and adrenalectomy. J Neuroendocrinol 18:655–661

Nolan LA, Kavanagh E, Lightman SL, Levy A (1998) Anterior pituitary cell population control: basal cell turnover and the effects of adrenalectomy and dexamethasone treatment. J Neuroendocrinol 10:207–215

Pevny LH, Nicolis SK (2010) Sox2 roles in neural stem cells. Int J Biochem Cell Biol 42:421–424

Raetzman LT, Ross SA, Cook S, Dunwoodie SL, Camper SA, Thomas PQ (2004) Developmental regulation of Notch signaling genes in the embryonic pituitary: Prop1 deficiency affects Notch2 expression. Dev Biol 265:329–340

Rando TA (2006) Stem cells, ageing and the quest for immortality. Nature 441:1080–1086

Rizzoti K, Akiyama H, Lovell-Badge R (2013) Mobilized adult pituitary stem cells contribute to endocrine regeneration in response to physiological demand. Cell Stem Cell 13:419–432

Roskams T (2006) Different types of liver progenitor cells and their niches. J Hepatol 45:1–4

Schepers AG, Snippert HJ, Stange DE, van den Born M, van Es JH, van de Wetering M, Clevers H (2012) Lineage tracing reveals Lgr5+ stem cell activity in mouse intestinal adenomas. Science 337:730–735

Schneider HJ, Aimaretti G, Kreitschmann-Andermahr I, Stalla GK, Ghigo E (2007) Hypopituitarism. Lancet 369:1461–1470

Sergeant G, Vankelecom H, Gremeaux L, Topal B (2009) Role of cancer stem cells in pancreatic ductal adenocarcinoma. Nat Rev Clin Oncol 6:580–586

Sierra A, Martin-Suarez S, Valcarcel-Martin R, Pascual-Brazo J, Aelvoet SA, Abiega O, Deudero JJ, Brewster AL, Bernales I, Anderson AE, Baekelandt V, Maletic-Savatic M, Encinas JM (2015) Neuronal hyperactivity accelerates depletion of neural stem cells and impairs hippocampal neurogenesis. Cell Stem Cell 216:488–503

Slack JM (2008) Origin of stem cells in organogenesis. Science 322:1498–1501

Solanas G, Batlle E (2011) Control of cell adhesion and compartmentalization in the intestinal epithelium. Exp Cell Res 317:2695–2701

Susa T, Kato T, Yoshida S, Yako H, Higuchi M, Kato Y (2012) Paired-related homeodomain proteins Prx1 and Prx2 are expressed in embryonic pituitary stem/progenitor cells and may be involved in the early stage of pituitary differentiation. J Neuroendocrinol 24:1201–1212

Takahashi K, Yamanaka S (2006) Induction of pluripotent stem cells from mouse embryonic and adult fibroblast cultures by defined factors. Cell 126:663–676

Tando Y, Fujiwara K, Yashiro T, Kikuchi M (2013) Localization of Notch signaling molecules and their effect on cellular proliferation in adult rat pituitary. Cell Tissue Res 351:511–519

Tanriverdi F, Schneider HJ, Aimaretti G, Masel BE, Casanueva FF, Kelestimur F (2015) Pituitary dysfunction after traumatic brain injury: a clinical and pathophysiological approach. Endocr Rev 36:305–342

Ueharu H, Higuchi M, Nishimura N, Yoshida S, Shibuya S, Sensui K, Kato T, Kato Y (2014) Expression of Krüppel-like factor 6, KLF6, in rat pituitary stem/progenitor cells and its regulation of the PRRX2 gene. J Reprod Dev 60:304–311

Van Aken MO, Lamberts SW (2005) Diagnosis and treatment of hypopituitarism: an update. Pituitary 8:183–191

Van den Broeck A, Vankelecom H, Van Delm W, Gremeaux L, Wouters J, Allemeersch J, Govaere O, Roskams T, Topal B (2013) Human pancreatic cancer contains a side population expressing cancer stem cell-associated and prognostic genes. PLoS One 8, e73968

Vankelecom H (2010) Pituitary stem/progenitor cells: embryonic players in the adult gland? Eur J Neurosci 32:2063–2081

Vankelecom H (2012) Pituitary stem cells drop their mask. Curr Stem Cell Res Ther 7:36–71

Vankelecom H, Chen J (2014) Pituitary stem cells: where do we stand? Mol Cell Endocrinol 385:2–17

Vankelecom H, Gremeaux L (2010) Stem cells in the pituitary gland: a burgeoning field. Gen Comp Endocrinol 166:478–488

Wabik A, Jones PH (2015) Switching roles: the functional plasticity of adult tissue stem cells. EMBO J 34:1164–1179

Wagers AJ, Conboy IM (2005) Cellular and molecular signatures of muscle regeneration: current concepts and controversies in adult myogenesis. Cell 122:659–667

Ward RD, Raetzman LT, Suh H, Stone BM, Nasonkin IO, Camper SA (2005) Role of PROP1 in pituitary gland growth. Mol Endocrinol 19:698–710

Willems C, Vankelecom H (2014) Pituitary cell differentiation from stem cells and other cells: towards restorative therapy for hypopituitarism? Regen Med 9:513–534

Willems C, Fu Q, Roose H, Mertens F, Cox B, Chen J, Vankelecom H (2016) Regeneration in the pituitary after cell-ablation injury: time-related aspects and molecular analysis. Endocrinology 157:705–21

Wouters J, Vankelecom H, van den Oord J (2009) Cancer stem cells in cutaneous melanoma. Expert Rev Dermatol 4:225–235

Wouters J, Stas M, Gremeaux L, Govaere O, Van den Broeck A, Maes H, Agostinis P, Roskams T, van den Oord JJ, Vankelecom H (2013) The human melanoma side population displays molecular and functional characteristics of enriched chemoresistance and tumorigenesis. PLoS One 8, e76550

Xu X, D'Hoker J, Stange G, Bonne S, De LN, Xiao X, Van de Casteele M, Mellitzer G, Ling Z, Pipeleers D, Bouwens L, Scharfmann R, Gradwohl G, Heimberg H (2008) Beta cells can be generated from endogenous progenitors in injured adult mouse pancreas. Cell 132:197–207

Xu Q, Yuan X, Tunici P, Liu G, Fan X, Xu M, Hu J, Hwang JY, Farkas DL, Black KL, Yu JS (2009) Isolation of tumour stem-like cells from benign tumours. Br J Cancer 101:303–311

Yin H, Price F, Rudnicki MA (2013) Satellite cells and the muscle stem cell niche. Physiol Rev 93:23–67

Yoshida S, Kato T, Susa T, Cai LY, Nakayama M, Kato Y (2009) PROP1 coexists with SOX2 and induces PIT1-commitment cells. Biochem Biophys Res Commun 385:11–15

Yoshida S, Kato T, Yako H, Susa T, Cai LY, Osuna M, Inoue K, Kato Y (2011) Significant quantitative and qualitative transition in pituitary stem/progenitor cells occurs during the postnatal development of the rat anterior pituitary. J Neuroendocrinol 23:933–943

Yoshida S, Kato T, Higuchi M, Chen M, Ueharu H, Nishimura N, Kato Y (2015) Localization of juxtacrine factor ephrin-B2 in pituitary stem/progenitor cell niches throughout life. Cell Tissue Res 359:755–766

Zhu X, Lin CR, Prefontaine GG, Tollkuhn J, Rosenfeld MG (2005) Genetic control of pituitary development and hypopituitarism. Curr Opin Genet Dev 15:332–340

Zhu X, Gleiberman AS, Rosenfeld MG (2007) Molecular physiology of pituitary development: signaling and transcriptional networks. Physiol Rev 87:933–963

Pituitary Stem Cells During Normal Physiology and Disease

Cynthia L. Andoniadou

Summary The homeostatic maintenance and functional modification of tissues require a combination of regulated proliferation and differentiation by somatic stem cells and more committed progenitors. Of relevance to regenerative medicine approaches, the endogenous stimulation of cell types for replenishment of damaged tissues requires an understanding of the signals that promote proliferation and direct appropriate differentiation to specialised cell types. We recently showed that pituitary stem cells expressing the transcription factor SOX2 are able to contribute to the generation of new hormone-producing cells during postnatal life. The signals controlling proliferation in the anterior pituitary are poorly understood and little is known about the influences supporting the choices between proliferation and quiescence among stem cells. The WNT signalling pathway is a major regulator of proliferation and influences stem cells in multiple tissues throughout the body as well as cancer stem cells in tumorigenesis. Forced up-regulation of the WNT pathway specifically in SOX2-positive pituitary stem cells by transgenic approaches in mouse stimulates a transient burst of proliferation, maintaining their uncommitted phenotype. These mutated stem cells subsequently induce tumorigenesis in a non-cell autonomous manner, as they promote proliferation of surrounding cell types through the secretion of paracrine factors. The studies presented here aim to provide insights into pituitary stem cell behaviour and their possible roles during disease states.

"If there were no regeneration there could be no life. If everything regenerated there would be no death. All organisms exist between these two extremes." Richard J. Goss, *Principles of Regeneration* (1969).

C.L. Andoniadou (✉)
Division of Craniofacial Development and Stem Cell Biology, King's College London, Floor 27 Tower Wing, Guy's Campus, London SE1 9RT, United Kingdom
e-mail: cynthia.andoniadou@kcl.ac.uk

© The Author(s) 2016
D. Pfaff, Y. Christen (eds.), *Stem Cells in Neuroendocrinology*, Research and Perspectives in Endocrine Interactions, DOI 10.1007/978-3-319-41603-8_8

103

Introduction

Over the last few years, compelling evidence has demonstrated the presence of adult somatic stem cells in the murine pituitary gland of mice. In this chapter I will summarise the critical in vitro and in vivo evidence demonstrating the presence and functional properties of these cells. In addition, I will highlight how pituitary stem cells can be involved in tumour formation, as has been shown for stem cell populations of other organs.

The regulation of stem cell populations is of interest to regenerative medicine and cancer therapy approaches. Being able to studying the behaviour of stem cells in their tissue niches can lead to a better understanding of how these behave and how they are regulated. An abnormal expansion or depletion of such populations may contribute to neoplasias or organ failure, respectively. In the case of the pituitary gland, this would manifest as hypopituitarism or pituitary tumours. Recently, we (Andoniadou/Martinez-Barbera labs) and other groups have provided evidence that a long-lived, tissue-specific population of undifferentiated progenitor/ stem cells exists within the anterior pituitary gland. Pituitary stem cells (PSCs) are undifferentiated and are able to generate cells of three main progenitor lineages, characterized by expression of the transcription factors, PIT1 (POU1F1), TPIT (TBX19) and SF1 (NR5A1), the expression of which is necessary for terminal differentiation into hormone-secreting cell types. PIT1-positive progenitors are the major lineage of the anterior pituitary and give rise to somatotrophs expressing growth hormone (GH), lactotrophs expressing prolactin (PRL) and thyrotrophs expressing thyroid stimulating hormone (TSH). Progenitors positive for TPIT give rise to ACTH-expressing adrenocorticotrophs and MSH-expressing melano-trophs (refer to Jacques Drouin 2016 for the transcriptional mechanisms regulating these fate choices). Finally, progenitors expressing SF1 cells give rise to gonado-trophs that express LH or FSH. All of these populations need to be precisely regulated to ensure appropriate homeostasis and adequate response for hormone secretion dependent on physiological demand, a process that is very dynamic throughout life.

To be considered PSCs, the cells need to demonstrate both self-renewal and differentiation in vivo. Initial studies pointing towards PTCs relied on in vitro approaches. These identified that there are cells in the postnatal pituitary gland that have the capacity to expand as colonies in culture, i.e., they have clonogenic capacity, demonstrating self-renewal in vitro. One population was characterized by the uptake of the fluorescent dipeptide AMCA and included cells expressing S100 calcium-binding protein B (S100β; Lepore et al. 2005), which have been described as folliculostellate cells of the anterior lobe (Vila-Porcile 1972). Cells among this population were able to form adherent colonies in culture. The second population was characterized by marker expression similar to stem cells of other tissues (*Sca1*, *Nanog* and *Oct4*) and could form floating spheres (Chen et al. 2005). In later studies, the same research group showed that there was enrichment in this population for the expression of SOX2, SOX9, CD44 and CD133 (Chen et al. 2009).

Additionally, these cells had active function of ABC transporters, which rendered them capable of effluxing the vital dye Hoechst 33342, leading to a discrete 'side population' during flow cytometry when analysing levels of expression of Hoechst. This is a typical property of many cell types that possess properties of stem or progenitor cells. This side population has since been described for pituitaries of other vertebrates (Chen et al. 2005, 2006; van Rijn et al. 2012). In time, additional markers of cells with this in vitro self-renewal capacity have been put forward to refine the characterisation of PSCs, including Nestin, PROP1, SOX9, GFRα2 and PRX1/2 (Gleiberman et al. 2008; Yoshida et al. 2009; Rizzoti et al. 2013; Garcia-Lavandeira et al. 2009; Higuchi et al. 2014). In the postnatal rodent gland in vivo, SOX2-positive cells displayed a high degree of overlap with other proposed stem cell markers such as SOX9 (Rizzoti et al. 2013), PROP1 (Yoshida et al. 2009, 2011), and PRX1/2 (Higuchi et al. 2014). We confirmed that SOX2-positive cells did not overlap with differentiation markers but we sought to determine if they were able to give rise to all the differentiated lineages in vivo.

The Lovell-Badge lab reported that floating spheres forming from anterior pituitary cells, were positive for SOX2 (Fauquier et al. 2008), in line with *Sox2* expression in the side population (Gremeaux et al. 2012). Our group utilised a knock-in mouse strain expressing enhanced yellow fluorescent protein (EYFP) from the SOX2 locus, allowing identification of SOX2-positive cells (Andoniadou et al. 2012). We isolated SOX2-EYFP-positive and -negative populations separately and plated these under conditions promoting the clonogenic expansion of single cells as adherent colonies. Only cells within the SOX2-positive fraction were capable of colony formation. This was also the case when clonogenic potential was assessed via the generation of floating spheres, as reported by the Lovell-Badge group (Rizzoti 2010). In the adherent cultures, time-lapse imaging of singe cells plated at clonal density confirmed that they gave rise to single colonies that contained multiple EYFP-positive cells, demonstrating self-renewal. Withdrawal of growth factors and prolonged culture in differentiation conditions promoted the expression of markers of the three main pituitary lineages and the expression of differentiation markers, as detected by qPCR (Andoniadou et al. 2012). Interestingly, only a small percentage of SOX2$^+$ cells were able to form colonies (up to 5 % depending on age), possibly reflecting heterogeneity within this population in terms of their potential; alternatively, the culture requirements allowed for expansion of a restricted subset of cells, where more cell types could have self-renewal capacity. In either case, the requirement for a better-defined combination of markers was clear. To address this issue, we investigated S100β as an additional marker of the pituitary stem cell population, as these had a significant overlap with SOX2$^+$ cells in vivo (Fauquier et al. 2008; Andoniadou et al. 2013). We purified this population from transgenic mice expressing S100β-GFP and found that plating GFP-positive and -negative cells resulted in the enrichment in colonies forming in the GFP-positive compartment. Since this property lies solely within the SOX2-positive cells in our assay (Andoniadou et al. 2013), pituitary stem cells are likely to be enriched within the double-positive population. Double SOX2/S100β are located along the marginal zone, the epithelium lining the remnants of Rathke's pouch lumen, and also in the parenchyma of the anterior pituitary, often in groups distributed amongst hormone-secreting cells. We isolated SOX2 cells expressing GFP from these two regions by microdissection, and demonstrated that their in vitro clonogenic potential did not differ between the two locations (Andoniadou et al. 2012). Recent studies from rat have revealed that double-positive SOX2$^+$/S100β^+ cells may be further refined through in vivo expression of the gene *Cxadr*, which codes for coxsackievirus and adenovirus receptor (CAR; Chen et al. 2013). Furthermore, expression of E-cadherin and the juxtacrine factor ephrin-B2 reportedly define SOX2$^+$/S100β^+/CAR$^+$ cells, both in the marginal epithelium and throughout the parenchyma (Chen et al. 2013; Yoshida et al. 2015). Analysing the side population, the Vankelecom group (2010) also reported enrichment in ephrin-B expression in this stem cell-rich compartment; together the data suggested that ephrin-B expression was a hallmark of the population containing PSCs.

The Long-Term Maintenance of the Anterior Pituitary

Until recently, evidence to support that cells in the pituitary could act as stem cells in vivo was lacking. This evidence has now been provided through genetic tracing of SOX2$^+$ cells, enabled by the generation of inducible mouse strains expressing CreERT2 under the regulation of the SOX2 promoter, where Cre recombinase is expressed in SOX2$^+$ cells but will not be active until the administration of tamoxifen, allowing temporal control of recombinase action (Andoniadou et al. 2013; Arnold et al. 2011). We lineage traced cells expressing *Sox2* both during gestation and postnatally (Andoniadou et al. 2013). Similarly, the Lovell-Badge group traced *Sox2*-expressing and *Sox9*-expressing cells from embryonic stages (Rizzoti et al. 2013). In all cases, these populations gave rise to all committed progenitor cell types (PIT1, TPIT, SF1) and hormone-secreting cells of the anterior lobe (GH, PRL, TSH, ACTH, LH/FSH). A similar capacity by both SOX2$^+$ and SOX9$^+$ populations is not surprising since *Sox2* and *Sox9* are co-expressed in the majority of cells from late embryonic stages (Rizzoti et al. 2013). What remains unknown is the extent of heterogeneity within this population. Our assay does not distinguish if a SOX2$^+$ cell capable of proliferation is multipotent or if there are several distinct oligopotent SOX2$^+$ sub-populations that collectively cover the different populations. It is, however, able to demonstrate that the population of SOX2$^+$ cells is long-lived and does not become depleted over time, something that would be expected of a transit-amplifying progenitor population. We activated CreERT2 in postnatal SOX2$^+$ cells, enabling expression of *R26R-EYFP*, and traced the cells for six months. We assessed the clonogenic potential of EYFP$^+$ cells at the end of this period, i.e., the SOX2$^+$ cells as well as their derivatives. We flow sorted for EYFP expression and cultured the positive and negative populations under adherent clonogenic conditions. Even after six months, the majority of cells with clonogenic potential (subpopulation of SOX2$^+$) resided within the EYFP$^+$ fraction, indicating either that SOX2 cells were long-lived, hence persisting long-term after initial labeling, or that the pool of SOX2$^+$ cells was maintained through self-renewal. Importantly, it excluded the possibility that the SOX2$^+$ pituitary stem cell pool became depleted. Immunofluorescence staining confirmed that, even after a year of tracing following tamoxifen administration, EYFP$^+$ cells included SOX2$^+$ uncommitted cells. The above experiments relied on repeated administration of a high dosage of tamoxifen to ensure that the majority of SOX2$^+$ cells were initially labelled. We have further analysed the behaviour of SOX2$^+$ cells by labelling sparse cells through low dosage administration. The above experiments demonstrate the presence of a long-lived population that retains pituitary stem cell properties throughout normal life. Ongoing efforts focus on the signals that regulate activity of this stem cell compartment and on how the potential of PSCs is maintained during life.

Stem Cells and Pituitary Tumours

Similar to organ-specific stem cells, analyses of many tumours and cancers have revealed the presence of multipotent cells, which are often thought to drive tumour formation. Many of the properties of 'cancer stem cells' (CSCs) are shared by normal tissue-specific stem cells: slow cycling status, self-renewal and differentiation capacity and even resistance to cytotoxic drugs. For some tumours, it has been shown that normal stem cells are transformed into CSCs when targeted to express oncogenic proteins. For example, intestinal crypt stem cells transform into CSCs when the WNT/β-catenin pathway is over-activated (Barker et al. 2009; Zhu et al. 2009). However, progenitor cells or even differentiated cells could give rise to cells fulfilling CSC criteria upon transformation (Valent et al. 2012; Clevers 2011). Several groups have reported the presence of putative CSCs in human pituitary adenomas and from mouse pituitary tumour models (Chen et al. 2014; Xu et al. 2009; Donangelo et al. 2014; Lloyd et al. 2013; Mertens et al. 2015; Orciani et al. 2015; van Rijn et al. 2013; Yunoue et al. 2011; Hosoyama et al. 2010). It remains unknown if CSCs arise from PSCs, although their properties can be similar; for example, both are capable of in vitro self-renewal and differentiation and PSCs are likely to be chemoresistant like CSCs, since PSCs are found within the 'side population' generated by dye efflux, utilising the same transporter properties as chemoresistance (Chen et al. 2009).

In our quest to determine if PSCs are transformed to CSCs following expression of an oncogenic protein, we uncovered a non-cell autonomous role for these cells in tumour formation through the expression of paracrine factors that promote tumour formation by a different cell population. We focused on mutations in *CTNNB1*, the gene encoding β-catenin; these mutations have been identified in numerous tumours and in the pituitary, identified in the majority of adamantinomatous craniopharyngioma (ACP) tumours (Buslei et al. 2005). ACPs are aggressive tumours with a tendency to infiltrate the brain, vascular structures and optic tracts (Muller 2014). They represent the most common pituitary tumour type in children and are mostly paediatric (Muller 2013). The mutations responsible for the generation of ACP activate the WNT/β-catenin pathway by preventing the degradation of β-catenin and resulting in its accumulation (Martinez-Barbera 2015). A hallmark of ACP is the presence of small cell clusters that strongly accumulate nucleo-cytoplasmic β-catenin, as revealed by immunohistochemistry (Buslei et al. 2007; Hofmann et al. 2006). We expressed this mutation embryonically throughout the developing pituitary primodium, from the early specification of Rathke's pouch using the *Hesx1-Cre* driver. *Hesx1*$^{Cre/+}$; *Ctnnb1*$^{lox(ex3)/+}$ animals develop tumours very similar to human ACP and contain the hallmark β-catenin-accumulating cell clusters that activate the WNT/β-catenin pathway (Gaston-Massuet et al. 2011). We assessed the mouse tumours for the presence of cells reminiscent of CSCs, which have self-renewal and differentiation properties in vitro, and found an increase of clonogenic cells, up to three times the number compared to normal pituitaries (Gaston-Massuet et al. 2011). These cells expressed markers of stem cells such as *Sox2*; however, immunofluorescence experiments showed that these predominantly co-localised within the β-catenin-accumulating clusters. This population of SOX2$^+$ cells was expanded compared to normal numbers. In an effort to identify if these cells acted as CSCs in this neoplasm to generate the tumour mass, we pursued a different approach. We expressed the oncogenic β-catenin specifically in SOX2$^+$ stem cells of postnatal pituitary glands using a mouse model where timing of expression of the oncogenic protein is dependent upon administration of tamoxifen (*Sox2*$^{CreERT2/+}$; *Ctnnb1*$^{lox(ex3)/+}$ mouse model). This approach resulted in tumours that were similar to human ACP and the previous embryonic model (Andoniadou et al. 2013). The advantage of this approach was that it allowed us to lineage-trace the fate of the PSCs carrying the oncogenic mutation and determine their contribution to the tumour mass. As done previously for normal PSC lineage tracing, SOX2$^+$ cells were targeted simultaneously to oncogenic β-catenin and EYFP, allowing identification of daughter cells derived from the mutated SOX2$^+$ PSCs (*Sox2*$^{CreERT2/+}$; *Ctnnb1*$^{lox(ex3)/+}$;*R26*$^{EYFP/+}$). These experiments confirmed that the typical cell clusters derived from SOX2$^+$ cells but, intriguingly, they revealed that the bulk of the tumour mass did not. Therefore, mutated SOX2$^+$ PSCs were not transformed into CSCs by oncogenic β-catenin; instead, the cell clusters that they generated were found to have the capacity to induce tumours through paracrine signalling (Andoniadou et al. 2013). Through isolating the cluster cells and performing gene expression analyses, we identified that they expressed a vast array of growth factors, chemokines and cytokines including members of the TGF, FGF and PDGF families of growth factors among many others. These cells could therefore act as signalling centres, likely changing the microenvironment and facilitating tumorigenesis (Andoniadou et al. 2012, 2013). An observation that was indicative of the potential of these cluster cells was that actively proliferating cells, as marked by immunofluorescence of Ki67, were readily detected in close proximity to the cell clusters in both mouse and human ACP (Gaston-Massuet et al. 2011). Several of the identified cytokines and growth factors have been shown to play a role in normal pituitary physiology as well as in pituitary adenomas (Arzt et al. 1999, 2009; Graciarena et al. 2004). Future work aims to reveal the mechanisms whereby cell clusters may induce paracrine cell transformation and promote tumour growth as well as the cell of origin of the ACP tumour mass. Taken together our data reveal that pituitary gland stem cells have the potential to contribute to tumorigenesis when mutated. In the case of ACP, they can instigate tumour formation in a non-cell autonomous manner, but this does not preclude the possibility of them acting as CSCs in other tumours.

Further research will better characterise PCS to reveal their defining features and potential as well as the mechanisms that regulate their activities, which can only enhance our ability to understand their possible role in disease and future regenerative medicine approaches, leading to more effective prognoses and treatments.

References

Andoniadou CL, Gaston-Massuet C, Reddy R, Schneider RP, Blasco MA, Le Tissier P, Jacques TS, Pevny LH, Dattani MT, Martinez-Barbera JP (2012) ntification of novel pathways involved in the pathogenesis of human adamantinomatous craniopharyngioma. Acta Neuropathol 124: 259–271

Andoniadou CL, Matsushima D, Mousavy Gharavy SN, Signore M, Mackintosh AI, Schaeffer M, Gaston-Massuet C, Mollard P, Jacques TS, Le Tissier P, Dattani MT, Pevny LH, Martinez-Barbera JP (2013) Sox2(+) stem/progenitor cells in the adult mouse pituitary support organ homeostasis and have tumor-inducing potential. Cell Stem Cell 13:433–445

Arnold K, Sarkar A, Yram MA, Polo JM, Bronson R, Sengupta S, Seandel M, Geijsen N, Hochedlinger K (2011) Sox2(+) adult stem and progenitor cells are important for tissue regeneration and survival of mice. Cell Stem Cell 9:317–329

Arzt E, Pereda MP, Castro CP, Pagotto U, Renner U, Stalla GK (1999) Pathophysiological role of the cytokine network in the anterior pituitary gland. Front Neuroendocrinol 20:71–95

Arzt E, Chesnokova V, Stalla GK, Melmed S (2009) Pituitary adenoma growth: a model for cellular senescence and cytokine action. Cell Cycle 8:677–678

Barker N, Ridgway RA, van Es JH, van de Wetering M, Begthel H, van den Born M, Danenberg E, Clarke AR, Sansom OJ, Clevers H (2009) Crypt stem cells as the cells-of-origin of intestinal cancer. Nature 457:608–611

Buslei R, Nolde M, Hofmann B, Meissner S, Eyupoglu IY, Siebzehnrubl F, Hahnen E, Kreutzer J, Fahlbusch R (2005) Common mutations of beta-catenin in adamantinomatous craniopharyngiomas but not in other tumours originating from the sellar region. Acta Neuropathol 109:589–597

Buslei R, Holsken A, Hofmann B, Kreutzer J, Siebzehnrubl F, Hans V, Oppel F, Buchfelder M, Fahlbusch R, Blümcke I (2007) Nuclear beta-catenin accumulation associates with epithelial morphogenesis in craniopharyngiomas. Acta Neuropathol 113:585–590

Chen J, Hersmus N, Van Duppen V, Van Duppen V, Caesens P, Denef C, Vankelecom H (2005) The adult pituitary contains a cell population displaying stem/progenitor cell and early embryonic characteristics. Endocrinology 146:3985–3998

Chen J, Crabbe A, Van Duppen V, Vankelecom H (2006) The notch signaling system is present in the postnatal pituitary: marked expression and regulatory activity in the newly discovered side population. Mol Endocrinol 20:3293–3307

Chen J, Gremeaux L, Fu Q, Liekens D, Van Laere S, Vankelecom H (2009) Pituitary progenitor cells tracked down by side population dissection. Stem Cells 27:1182–1195

Chen M, Kato T, Higuchi M, Yoshida S, Yako H, Kanno N, Kato Y (2013) Coxsackievirus and adenovirus receptor-positive cells compose the putative stem/progenitor cell niches in the marginal cell layer and parenchyma of the rat anterior pituitary. Cell Tissue Res 354:823–836

Chen L, Ye H, Wang X, Tang X, Mao Y, Zhao Y, Wu Z, Mao XO, Xie L, Jin K, Yao Y (2014) Evidence of brain tumor stem progenitor-like cells with low proliferative capacity in human benign pituitary adenoma. Cancer Lett 349:61–66

Clevers H (2011) The cancer stem cell: premises, promises and challenges. Nat Med 17:313–319

Donangelo I, Ren SG, Eigler T, Svendsen C, Melmed S (2014) Sca1(+) murine pituitary adenoma cells show tumor-growth advantage. Endocr Relat Cancer 21:203–216

Drouin J (2016) Epigenetic mechanisms of pituitary cell fate specification. In: Pfaff D, Christen Y (eds) Stem cells in neuroendocrinology. Springer, Heidelberg

Fauquier T, Rizzoti K, Dattani M, Lovell-Badge R, Robinson IC (2008) SOX2-expressing progenitor cells generate all of the major cell types in the adult mouse pituitary gland. Proc Natl Acad Sci U S A 105:2907–2912

Garcia-Lavandeira M, Quereda V, Flores I, Saez C, Diaz-Rodriguez E, Japon MA, Ryan AK, Blasco MA, Dieguez C, Malumbres M, Alvarez CV (2009) A GRFa2/Prop1/stem (GPS) cell niche in the pituitary. PLoS One 4, e4815

Gaston-Massuet C, Andoniadou CL, Signore M, Jayakody SA, Charolidi N, Kyeyune R, Vernay B, Jacques TS, Taketo MM, Le Tissier P, Dattani MT, Martinez-Barbera JP (2011) Increased Wingless (Wnt) signaling in pituitary progenitor/stem cells gives rise to pituitary tumors in mice and humans. Proc Natl Acad Sci U S A 108:11482–11487

Gleiberman AS, Michurina T, Encinas JM, Roig JL, Krasnov P, Balordi F, Fishell G, Rosenfeld MG, Enikolopov G (2008) Genetic approaches identify adult pituitary stem cells. Proc Natl Acad Sci U S A 105:6332–6337

Graciarena M, Carbia-Nagashima A, Onofri C, Perez-Castro C, Giacomini D, Renner U, Stalla GK, Artz E (2004) Involvement of the gp130 cytokine transducer in MtT/S pituitary somatotroph tumour development in an autocrine-paracrine model. Eur J Endocrinol 151:595–604

Gremeaux L, Fu Q, Chen J, Vankelekom H (2012) Activated phenotype of the pituitary stem/progenitor cell compartment during the early-postnatal maturation phase of the gland. Stem Cells Dev 21:801–813

Higuchi M, Yoshida S, Ueharu H, Chen M, Kato T, Kato Y (2014) PRRX1 and PRRX2 distinctively participate in pituitary organogenesis and a cell-supply system. Cell Tissue Res 357:323–335

Hofmann BM, Kreutzer J, Saeger W, Buchfelder M, Blümcke I, Fahlbusch R, Buslei R (2006) Nuclear beta-catenin accumulation as reliable marker for the differentiation between cystic craniopharyngiomas and rathke cleft cysts: a clinico-pathologic approach. Am J Surg Pathol 30:1595–1603

Hosoyama T, Nishijo K, Garcia MM, Schaffer BS, Ohshima-Hosoyama S, Prajapati SI, Davis MD, Grant WF, Scheithauer BW, Marks DL, Rubin BP, Keller C (2010) A postnatal Pax7 progenitor gives rise to pituitary adenomas. Genes Cancer 1:388–402

Lepore DA, Roeszler K, Wagner J, Ross SA, Bauer K, Thomas PQ (2005) Identification and enrichment of colony-forming cells from the adult murine pituitary. Exp Cell Res 308:166–176

Lloyd RV, Hardin H, Montemayor-Garcia C, Rotondo F, Syro LV, Horvath E, Kovacs K (2013) Stem cells and cancer stem-like cells in endocrine tissues. Endocr Pathol 24:1–10

Martinez-Barbera JP (2015) Molecular and cellular pathogenesis of adamantinomatous cranio-pharyngioma. Neuropathol Appl Neurobiol 41:721–732

Mertens FM, Gremeaux L, Chen J et al (2015) Pituitary tumors contain a side population with tumor stem cell-associated characteristics. Endocr Relat Cancer 22:481–504

Muller HL (2013) Paediatrics: surgical strategy and quality of life in craniopharyngioma. Nat Rev Endocrinol 9:447–449

Muller HL (2014) Craniopharyngioma. Endocr Rev 35(3):513–43. doi:10.1210/er.2013-1115

Orciani M, Davis S, Appolloni G et al (2015) Isolation and characterization of progenitor mesen-chymal cells in human pituitary tumors. Cancer Gene Ther 22:9–16

Rizzoti K (2010) Adult pituitary progenitors/stem cells: from in vitro characterization to in vivo function. Eur J Neurosci 32:2053–2062

Rizzoti K, Akiyama H, Lovell-Badge R (2013) Mobilized adult pituitary stem cells contribute to endocrine regeneration in response to physiological demand. Cell Stem Cell 13:419–432

Valent P, Bonnet D, De Maria R, Lapidot T, Copland M, Melo JV, Chomienne C, Ishikawa F, Schuringa JJ, Stassi G, Huntly B, Herrmann H, Soulier J, Roesch A, Schuurhuis GJ, Wöhrer S, Arock M, Zuber J, Cerny-Reiterer S, Johnsen HE, Andreeff M, Eaves C (2012) Cancer stem cell definitions and terminology: the devil is in the details. Nat Rev Cancer 12:767–775

van Rijn SJ, Gremeaux L, Riemers FM, Brinkhof B, Vankelecom H, Penning LC, Meij BP (2012) Identification and characterisation of side population cells in the canine pituitary gland. Vet J 192:476–482

van Rijn SJ, Tryfonidou MA, Hanson JM, Penning LC, Meij BP (2013) Stem cells in the canine pituitary gland and in pituitary adenomas. Vet Quart 33:217–224

Vankelecom H (2010) Pituitary stem/progenitor cells: embryonic players in the adult gland? Eur J Neurosci 32:2063–2081

Vila-Porcile E (1972) The network of the folliculo-stellate cells and the follicles of the adeno-hypophysis in the rat (pars distalis). Z Zellforsch Mikrosk Ana 129:328–369

Xu Q, Yuan X, Tunici P, Liu G, Fan X, Xu M, Hu J, Hwang JY, Farkas DL, Black KL, Yu JS (2009) Isolation of tumour stem-like cells from benign tumours. Br J Cancer 101:303–311

Yoshida S, Kato T, Susa T, Cai LY, Nakayama M, Kato Y (2009) PROP1 coexists with SOX2 and induces PIT1-commitment cells. Biochem Biophys Res Commun 385:11–15

Yoshida S, Kato T, Yako H, Susa T, Cai LY, Osuna M, Inoue K, Kato Y (2011) Significant quantitative and qualitative transition in pituitary stem / progenitor cells occurs during the postnatal development of the rat anterior pituitary. J Neuroendocrinol 23:933–943

Yoshida S, Kato T, Higuchi M, Ueharu H, Nishimura N, Kato Y (2015) Localization of juxtacrine factor ephrin-B2 in pituitary stem/progenitor cell niches throughout life. Cell Tissue Res 359: 755–766

Yunoue S, Arita K, Kawano H, Uchida H, Tokimura H, Hirano H (2011) Identification of CD133+ cells in pituitary adenomas. Neuroendocrinology 94:302–312

Zhu L, Gibson P, Currle DS, Tong Y, Richardson RJ, Bayazitov IT, Poppleton H, Zakharenko S, Ellison DW, Gilbertson RJ (2009) Prominin 1 marks intestinal stem cells that are susceptible to neoplastic transformation. Nature 457:603–607

Epigenetic Mechanisms of Pituitary Cell Fate Specification

Jacques Drouin

Abstract Pituitary progenitor or stem cells present in the pituitary primordium during development are the source of hormone-producing cells of the adult pituitary. These stem cells are maintained in the adult tissue and they can be recruited to maintain or replenish differentiated pituitary cells. We currently have only limited insight into the mechanisms that trigger progenitor engagement into one or the other pituitary differentiation pathway. While transcription factors that drive terminal differentiation have been identified for different lineages, current evidence suggests that initial engagement of progenitors into differentiation may be due to earlier-acting factors. One such factor expressed at the transition between progenitor and differentiated state was identified in the intermediate lobe; this factor, Pax7, exerts its action through a pioneer factor activity. Pioneer transcription factors have the unique ability to bind target sequences in compacted chromatin and to initiate chromatin "opening" for recruitment of other transcription factors. Pax7 accomplishes this process on about 2500 enhancers genome-wide, allowing for Tpit recruitment at a subset for implementation of the melanotrope-specific program of gene expression. Current knowledge about this process is reviewed here, together with a discussion of future challenges in order to understand the unique properties of pioneer transcription factor action and cell reprogramming through chromatin remodelling.

Introduction

The last decade was rich in surprising discoveries about the organization and function of the pituitary. Indeed, we realized that pituitary cells are organized in a series of intimately associated homotypic cell networks that serve to coordinate tissue response (Mollard et al. 2012). Further, the pituitary, like many other tissues,

J. Drouin (✉)
Laboratoire de Génétique Moléculaire, Institut de Recherches Cliniques de Montréal (IRCM), 110, avenue des Pins Ouest, Montréal, QC, Canada, H2W 1R7
e-mail: jacques.drouin@ircm.qc.ca

© The Author(s) 2016
D. Pfaff, Y. Christen (eds.), *Stem Cells in Neuroendocrinology*, Research and Perspectives in Endocrine Interactions, DOI 10.1007/978-3-319-41603-8_9

113

was found to contain a population of stem or progenitor cells that are maintained in adult tissues (Rizzoti 2015); this population of pituitary progenitors itself forms a homotypic network that is primarily organized around the cleft between the intermediate and anterior pituitary (Gremeaux et al. 2012). Signals that activate the progenitors to either proliferate or differentiate are still being investigated, but clearly these mechanisms provide a unique opportunity to think about new therapeutic perspectives to treat pituitary hormone deficiencies.

To capitalize on the properties of pituitary stem or progenitor cells for any kind of replacement therapy requires that we understand the unique properties of these cells and, most importantly, the mechanisms that engage stem cells into differentiation pathways. Pituitary progenitors can differentiate into each of the pituitary lineages (Fauquier et al. 2008); the challenge is then to understand the precise requirements for differentiation into each lineage. While a number of critical transcription factors have been identified for their role in terminal differentiation of most lineages (Rizzoti 2015), such as Pit1, Tpit and SF1, the initial events of entry into differentiation are the ones that we least understand presently.

This review will focus on one particular factor that appears to represent such an initial event for engagement into the differentiation pathway, Pax7, which selects intermediate lobe identity. Pax7 is unique among the transcription factors presently identified for a role in pituitary cell differentiation in that it possesses pioneer activity (Budry et al. 2012). Few transcription factors have the pioneer ability to bind "closed" or compacted chromatin and to trigger chromatin remodeling of regulatory sequences, thus opening not only chromatin structure but, most importantly, a new program of gene expression (Iwafuchi-Doi and Zaret 2014).

Escaping Stemness

Stem cells, whether pluripotent or tissue-specific, have unique properties with regards to basic cell physiology compared to differentiated cells (De Los Angeles et al. 2015). The engagement of stem cells into a differentiation pathway thus implies an important switch for many cellular functions in addition to the gain of expression for the unique genetic program of the chosen differentiation path. For example, this can include a switch in energy metabolism (Kohli and Passegue 2014); for pituitary progenitors, it was shown that the control of cell cycle re-entry is very different in progenitors compared to differentiated cells (Bilodeau et al. 2009). Indeed during normal mouse development, pituitary progenitors marked by expression of Sox2 (Fauquier et al. 2008) exit the cell cycle under the action of the Cip-Kip inhibitor $p57^{Kip2}$ whereas, upon differentiation, the role of keeper of cell cycle re-entry is taken over by the related $p27^{Kip1}$ (Fig. 1a). Although replacing one inhibitor by a related one may appear as switching between redundant regulators, the situation is likely more complex, since knockout mice for $p57^{Kip2}$ do eventually switch-on expression of $p27^{Kip1}$ with rescue of some differentiated cells; however, in the delayed progenitor compartment where this switch has not

A) Normal development

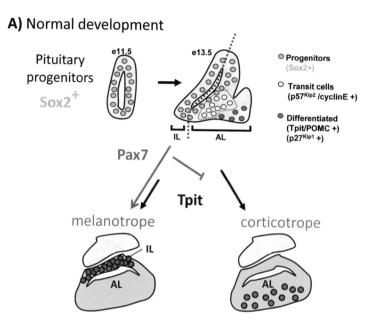

B) Gain-of-function – **Increased active enhancer repertoire**

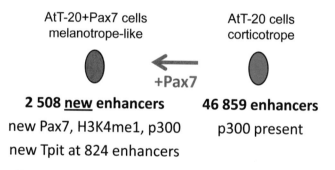

Fig. 1 Role of Pax7 in pituitary development. (**a**) The role of Pax7 in normal pituitary development is depicted in the context of the engagement of Sox2-positive pituitary progenitors into differentiation pathways. Pituitary progenitors are present throughout the pituitary primordium, Rathke's pouch, and the early pituitary depicted here at day e11.5 of mouse embryonic development. The Sox2-positive progenitors (*green*) proliferate and, starting at e12, a group of Sox2-negative, $p57^{Kip2}$- and cyclinE-positive cells (*white*) appears on the ventral side of the pituitary cleft: these cells have features of transitory precursors. The first differentiated cells appear on the ventral surface of the developing anterior lobe: these are Tpit-positive corticotropes (*blue*) that no longer express $p57^{Kip2}$ and CyclinE but rather $p27^{Kip1}$. Starting at e14, intermediate lobe (IL) cells express Pax7 followed by Tpit and later POMC. Anterior lobe (AL) corticotropes never express Pax7 and differentiate following expression of Tpit. (PL is posterior lobe). (**b**) Gain-of-function for Pax7 performed in corticotrope AtT-20 cells reveals the pioneer transcription factor activity of Pax7. The expression of Pax7 in AtT-20 cells reprograms the cells to become melanotrope-like (Budry et al. 2012). The action of Pax7 on the AtT-20 epigenome results in activation of 2508 new enhancers that have all the expected chromatin hallmarks; 824 of them can then recruit Tpit

occurred, massive apoptosis occurs, which is suggestive of a critical role for p57^{Kip2} in leaving the stem status.

Little is known about this switch from stem to differentiated state but, during development, it appears as sequential steps that are separated by a transitory period/ state where cells have lost markers of stemness such as Sox2 but not yet gained markers of differentiation such as expression of terminal differentiation factors like Pit1 and Tpit (Bilodeau et al. 2009). Cells undergoing this transitory period have been highlighted in the developing anterior pituitary but effectors of this transition remain to be identified. The big question is thus: what is happening to the genetic program of these transitory precursors? Are they being pre-programmed for leaving the stem state and for entry into one or the other pituitary differentiation pathway? Is their epigenome being set for establishment of a differentiation program?

In the intermediate pituitary, a transcription factor essential for intermediate lobe identity and melanotrope differentiation was identified and it appears to act within this gap between stem and differentiation states. Indeed, the transcription factor Pax7 is expressed immediately after extinction of Sox2, such that a small number of co-expressing cells can be visualized in the developing mouse pituitary (Budry et al. 2012). It is only about a half-day later that the driver of terminal differentiation, Tpit, is expressed in Pax7-positive cells, leading to expression of melanotrope genes (Fig. 1a).

Pax7 thus has the hallmark of a factor that would engage progenitors into a differentiation pathway and prepare transitory precursors for implementation of the terminal differentiation program that is itself directed by the Tbox factor Tpit. A factor with similar properties has yet to be discovered in the anterior pituitary.

Intermediate Pituitary Identity

The intermediate lobe melanotropes are one of the two pituitary lineages that express the single copy *POMC* gene. Terminal differentiation of both POMC lineages, the melanotropes and the corticotropes of the anterior lobe, requires the action of the Tbox transcription factor Tpit (Lamolet et al. 2001; Pulichino et al. 2003). Consistent with this role of Tpit as driver of terminal differentiation, Tpit is required for transcription of the *POMC* gene in both lineages and it acts genome-wide on an extensive program of gene expression (Langlais et al. 2011). Obviously by playing this role for both POMC lineages, Tpit cannot explain the unique features of melanotrope compared to corticotropes, the two lineages fulfilling entirely different biological functions and, accordingly, being regulated by very different signals.

The quest for transcription factor(s) that may account for the unique program of each lineage led to identification of Pax7 for its unique expression in the intermediate lobe melanotropes (Budry et al. 2012). The differentiation of most cell lineages requires the combinatorial action of many transcription factors that together control cell-specific gene expression. While it is in this context that Pax7 was initially investigated, it rapidly became clear that Pax7 has far more reaching actions on the epigenome and that it reprograms intermediate lobe cells for their unique purpose. Hence in *Pax7* knockout pituitaries, intermediate lobe cells still differentiate under the action of Tpit but, being deprived of Pax7, the cells switch fate and differentiate into corticotropes (Budry et al. 2012). This process is exemplified by the failure to activate melanotrope-specific genes such as the *PC2* gene, which encodes the protein convertase responsible for cleaving ACTH into αMSH, and also by repression of corticotrope-specific genes such as those for CRH and vasopressin receptors. This situation is very different in comparison to the *Tpit* knockout intermediate lobe, where cells fail to differentiate into either melanotrope or corticotrope but stay, for the most part, in a limbo between stem and differentiated states; for example, these cells co-express the two cell cycle inhibitors $p57^{Kip2}$ and $p27^{Kip1}$, a situation that is never observed normally. Thus, Pax7 itself does not drive expression of the differentiation program but rather sets the stage for Tpit action.

Pax7 Opens a New Enhancer Repertoire

The unique property of Pax7 to act as a pioneer transcription factor allows it to pre-program intermediate lobe precursors towards the melanotrope identity that will be later implemented by Tpit (Budry et al. 2012). The picture that is emerging with regards to pioneer factor action is still fragmentary, but the few pioneers that have been characterized may operate at different levels to remodel chromatin. Whereas the pluripotency factors Sox2, Oct4 and Klf4 initiate chromatin remodeling over large spans of the genome (Soufi et al. 2012), factors like FoxA (Cirillo et al. 2002), C/EBPα (van Oevelen et al. 2015), Ascl1(Wapinski et al. 2013) and Pax7 that are involved in specific-tissue programming act primarily on the enhancer repertoire by both locally opening chromatin at some enhancers and closing others for activity. The opening or priming of a new enhancer repertoire will de facto implement the possibility for a new program of gene expression and, indeed, this is what is observed in pituitary cells following Pax7 action: Tpit now gains access to a completely new set of enhancers (Fig. 1b). In gain-of-function experiments performed in AtT-20 cells, Pax7 triggered the local appearance of active chromatin marks at ~2500 enhancers and, of those, 824 became new targets of Tpit. Genome-wide, Pax7 targeted an enhancer repertoire that largely (73%) overlapped the Tpit target repertoire (Budry et al. 2012). The *PC2* (*PCSK2*) locus provides a good example of this: indeed in normal corticotrope AtT-20 cells, this gene is inactive and its enhancer bears no mark of active chromatin. Following Pax7 action, an enhancer located 146Kb upstream of the *PC2* transcription start site became active, as revealed by a variety of chromatin marks. Indeed, the enhancer acquired a bimodal distribution of histone H3K4me1 and the enhancer DNA became accessible where it was depleted of nucleosome. Also, the general co-activator p300 was recruited to the enhancer together with Tpit (Fig. 1b). This evolutionarily conserved enhancer thus has all the hallmarks of a Pax7- and Tpit-dependent enhancer (Budry et al. 2012).

In contrast to the pluripotency factors that have large-scale effects on chromatin organization, the action of Pax7 is restricted to a subset of enhancer domains. It is interesting that, in the AtT-20 cell gain-of-function experiments, Pax7 also partially repressed expression of corticotrope-specific genes and this was accompanied by a quantitative decrease in chromatin marks. However, the extinction of corticotrope-specific enhancers was at best partial in this model. Nonetheless, these observations indicate that Pax7 has the ability to not only activate a subset of enhancers but also to repress others, a property that is shared with other pioneer factors (Watts et al. 2011; Wang et al. 2015).

The Essence of Pioneering

The critical property of pioneer factors is that they can access their target DNA sequence in so-called "closed" or compacted heterochromatin. This unique property was well illustrated for Pax7 in comparison to Tpit at the *PC2* gene -146Kb enhancer, where Tpit could not gain access to its well-conserved palindromic target sequence if Pax7 had not previously directed local chromatin remodeling of the enhancer (Budry et al. 2012). How pioneer factors gain access to their sites in "closed" chromatin remains largely open to question. While some target binding sites may find themselves exposed on the surface of nucleosomal DNA, the random possibility of such positioning argues against this sole mechanism in view of the high selectivity exerted by pioneer factors. Some pioneers such as FoxA may initially bind its targets with decreased specificity to scan the genome and then bind more firmly at pioneering sites (Soufi et al. 2015), but this still does not explain why a specific subset of sites is selected. The winged helix DNA binding domain of FoxA factors may mimic the structure of the histone H1 linker that interacts with DNA to stabilize interaction with nucleosomal DNA (Cirillo and Zaret 2007), but this model will not apply to all pioneers such as for Pax7. Some factors interact with core histones, which may stabilize their association with chromatin (Cirillo et al. 2002; Fiedler et al. 2008). Other factors like the Sox family interact with the minor groove of DNA and that may facilitate recognition of target sequences on nucleosomes, but again, this property is not relevant for all pioneers.

For the melanotrope-specific enhancers pioneered by Pax7, the analysis of enhancers that contain only one sequence motif for Pax7 binding showed that the group of enhancers that have a so-called composite motif was specifically associated with pioneering sites (Budry et al. 2012). This composite target site contained each of the DNA sequences corresponding to the two DNA binding domains of Pax7, the paired and the homeo domains. In contrast, the enhancers that contained only the paired or homeodomain target sequence were primarily associated with enhancers that were already accessible and targeted by Pax7 for transcriptional activation without pioneering (Budry et al. 2012). Since the composite site was longer than either paired or homeodomain target sites, it is possible that higher affinity for this site may contribute to the ability to pioneer chromatin remodeling. Alternatively, it may also be that Pax7 interaction with the composite site involves a unique conformation of the Pax7 DNA binding domains that is itself required to initiate pioneering events.

Indeed, a unique conformation of the pioneer may be needed to recruit chromatin remodeling complexes that initiate the replacement of modified histone from repressive to active forms of histones. The sequence of biochemical changes required to establish a new stable chromatin environment at active enhancers remains poorly defined. In some systems, changing the chromatin environment requires passage through DNA replication, hence stripping of chromatin to re-establish a new chromatin environment (MacAlpine and Almouzni 2013; Nashun et al. 2015). It is an open question whether passage through DNA replication is needed for all pioneers (Iwafuchi-Doi and Zaret 2014), but for Pax7, the activation of target genes that require pioneering is far slower than for transcriptional activation of enhancers/genes that are already in an open active chromatin conformation. Hence, the action of Pax7 on chromatin remodeling may require passage through DNA replication.

Succeeding at Multiple Choices or Avoiding Cellular Schizophrenia

Pax7 is an interesting study case for pioneering. Indeed, we have shown a critical role for Pax7 in the establishment of intermediate pituitary identity and setting-up of the melanotrope program of gene expression, but the same Pax7 is also critical for the myogenic program of gene expression, where it is expressed at the transition between progenitor and differentiated skeletal muscles (Buckingham and Rigby 2014). These two entirely different functions of Pax7 are not compatible with each other and, despite the ability of Pax7 to recognize its target DNA sites in heterochromatin, it would need to distinguish the pituitary versus muscle targets in order to appropriately play its role in each tissue. How is this achieved? There is currently no clue on how this discrimination is exerted! The myogenic targets of Pax7 are not bound by Pax7 in pituitary cells, despite the fact that a large group of Pax7 target sites present in heterochromatin (marked with high levels of histone H3K9me3, the hallmark of heterochromatin) are bound by Pax7 in pituitary cells. Unexplainably, the myogenic targets of Pax7 exhibit lower levels of the repressive H3K9me3 in the same pituitary cells and yet are not bound by Pax7. Without providing an explanation, these observations nonetheless clearly indicate that we have much to learn about the nature of so-called "closed" chromatin and that there likely are many flavors of heterochromatin that remain to be defined.

Future Challenges

The above discussion has illustrated how pioneer transcription factors are important to implement the developmental program through their unique properties to (1) recognize their DNA targets on "closed" chromatin and (2) initiate chromatin remodeling either in a localized fashion at specific enhancers or more broadly on large genomic domains. It is critical to understand these processes in order to master cellular reprogramming for therapeutic uses.

In the context of pituitary development, we have reason to believe that there may be one or more pioneer factors that establish competence towards anterior pituitary lineages, and we would expect these factors to exert their critical function during the developmental window when fetal pituitary progenitors have lost expression of stem cell markers such as Sox2 but not yet gained expression of terminal differentiation drivers such Tpit or Pit1. Is there one pioneer for anterior pituitary identity that allows differentiation toward each of the five anterior lobe lineages or are there two for each of the related subgroups of lineages, the gonadotrope and corticotrope subgroup and the Pit-dependent subgroup? These are important questions that demand investigation.

References

Bilodeau S, Roussel-Gervais A, Drouin J (2009) Distinct developmental roles of cell cycle inhibitors p57Kip2 and p27Kip1 distinguish pituitary progenitor cell cycle exit from cell cycle re-entry of differentiated cells. Mol Cell Biol 29:1895–1908

Buckingham M, Rigby PW (2014) Gene regulatory networks and transcriptional mechanisms that control myogenesis. Dev Cell 28:225–238

Budry L, Balsalobre A, Gauthier Y, Khetchoumian K, L'honore A, Vallette S, Brue T, Figarella-Branger D, Meij B, Drouin J (2012) The selector gene Pax7 dictates alternate pituitary cell fates through its pioneer action on chromatin remodeling. Genes Dev 26:2299–2310

Cirillo LA, Zaret KS (2007) Specific interactions of the wing domains of FOXA1 transcription factor with DNA. J Mol Biol 366:720–724

Cirillo LA, Lin FR, Cuesta I, Friedman D, Jarnik M, Zaret KS (2002) Opening of compacted chromatin by early developmental transcription factors HNF3 (FoxA) and GATA-4. Mol Cell 9:279–289

De Los AA, Ferrari F, Xi R, Fujiwara Y, Benvenisty N, Deng H, Hochedlinger K, Jaenisch R, Lee S, Leitch HG, Lensch MW, Lujan E, Pei D, Rossant J, Wernig M, Park PJ, Daley GQ (2015) Hallmarks of pluripotency. Nature 525:469–478

Fauquier T, Rizzoti K, Dattani M, Lovell-Badge R, Robinson IC (2008) SOX2-expressing progenitor cells generate all of the major cell types in the adult mouse pituitary gland. Proc Natl Acad Sci USA 105:2907–2912

Fiedler M, Sanchez-Barrena MJ, Nekrasov M, Mieszczanek J, Rybin V, Muller J, Evans P, Bienz M (2008) Decoding of methylated histone H3 tail by the Pygo-BCL9 Wnt signaling complex. Mol Cell 30:507–518

Gremeaux L, Fu Q, Chen J, Vankelecom H (2012) Activated phenotype of the pituitary stem/progenitor cell compartment during the early-postnatal maturation phase of the gland. Stem Cells Dev 21:801–813

Iwafuchi-Doi M, Zaret KS (2014) Pioneer transcription factors in cell reprogramming. Genes Dev 28:2679–2692

Kohli L, Passegue E (2014) Surviving change: the metabolic journey of hematopoietic stem cells. Trends Cell Biol 24:479–487

Lamolet B, Pulichino AM, Lamonerie T, Gauthier Y, Brue T, Enjalbert A, Drouin J (2001) A pituitary cell-restricted T-box factor, Tpit, activates POMC transcription in cooperation with Pitx homeoproteins. Cell 104:849–859

Langlais D, Couture C, Sylvain-Drolet G, Drouin J (2011) A pituitary-specific enhancer of the POMC gene with preferential activity in corticotrope cells. Mol Endocrinol 25:348–359

MacAlpine DM, Almouzni G (2013) Chromatin and DNA replication. Cold Spring Harbor Perspec Biol 5:a010207

Mollard P, Hodson DJ, Lafont C, Rizzoti K, Drouin J (2012) A tridimensional view of pituitary development and function. Trends Endocrinol Metab 23:261–269

Nashun B, Hill PW, Hajkova P (2015) Reprogramming of cell fate: epigenetic memory and the erasure of memories past. Embo J 34:1296–1308

Pulichino AM, Vallette-Kasic S, Tsai JP, Couture C, Gauthier Y, Drouin J (2003) Tpit determines alternate fates during pituitary cell differentiation. Genes Dev 17:738–747

Rizzoti K (2015) Genetic regulation of murine pituitary development. J Mol Endocrinol 54: R55–73

Soufi A, Donahue G, Zaret KS (2012) Facilitators and impediments of the pluripotency reprogramming factors' initial engagement with the genome. Cell 151:994–1004

Soufi A, Garcia MF, Jaroszewicz A, Osman N, Pellegrini M, Zaret KS (2015) Pioneer transcription factors target partial DNA motifs on nucleosomes to initiate reprogramming. Cell 161:555–568

van Oevelen C, Collombet S, Vicent G, Hoogenkamp M, Lepoivre C, Badeaux A, Bussmann L, Sardina JL, Thieffry D, Beato M, Shi Y, Bonifer C, Graf T (2015) C/EBPalpha activates pre-existing and de novo macrophage enhancers during induced pre-B cell transdifferentiation and myelopoiesis. Stem Cell Rep 5:232–247

Wang A, Yue F, Li Y, Xie R, Harper T, Patel NA, Muth K, Palmer J, Qiu Y, Wang J, Lam DK, Raum JC, Stoffers DA, Ren B, Sander M (2015) Epigenetic priming of enhancers predicts developmental competence of hESC-derived endodermal lineage intermediates. Cell Stem Cell 16:386–399

Wapinski OL, Vierbuchen T, Qu K, Lee QY, Chanda S, Fuentes DR, Giresi PG, Ng YH, Marro S, Neff NF, Drechsel D, Martynoga B, Castro DS, Webb AE, Sudhof TC, Brunet A, Guillemot F, Chang HY, Wernig M (2013) Hierarchical mechanisms for direct reprogramming of fibroblasts to neurons. Cell 155:621–635

Watts JA, Zhang C, Klein-Szanto AJ, Kormish JD, Fu J, Zhang MQ, Zaret KS (2011) Study of FoxA pioneer factor at silent genes reveals Rfx-repressed enhancer at Cdx2 and a potential indicator of esophageal adenocarcinoma development. PLoS genetics 7, e1002277

Advances in Stem Cells Biology: New Approaches to Understand Depression

A. Borsini and P. A. Zunszain

Abstract Depression is a highly prevalent complex neuropsychiatric disorder, which ranks first among all mental and neurological disorders as a contributor to the global burden of disease. However, available treatments are still far from ideal, for their specificity as well as their efficacy. This situation can now be improved by the increasing availability of stem cells, which allows the development of in vitro human neural systems to model the brain. These models complement observations from animal models and patients with depression, allowing for a better understanding of the complexity of this psychiatric illness and potential treatments. Cells derived from the olfactory neuroepithelium, multipotent fetal hippocampal progenitor cells (HPCs) and human induced pluripotent stem cells (iPSCs) have shown promising leads. Using HPCs and iPSC-derived forebrain neurons, we managed to provide further insights into the action of drugs with antidepressant action as well as on molecular mechanisms underlying the effect of stress and inflammation, both linked to the pathophysiology of depression. Particular attention has been paid to the complex pathways by which the immune and stress systems differently determine the final developmental fate of HPCs and the synaptic plasticity of iPSCs. The combination of accessibility and validity of the available stem cells models will allow further work to increase our insights into the biology of depression and support the identification of novel therapeutic targets.

Introduction

How Can We Best Study Depression?

Depression ranks first among all mental and neurological disorders as a contributor to the global burden of disease and causes a heavy load on patients and their

A. Borsini • P.A. Zunszain (✉)
Department of Psychological Medicine, Section of Stress, Psychiatry and Immunology, King's College London, Institute of Psychiatry, Psychology and Neuroscience, London, UK
e-mail: patricia.zunszain@kcl.ac.uk

© The Author(s) 2016
D. Pfaff, Y. Christen (eds.), *Stem Cells in Neuroendocrinology*, Research and Perspectives in Endocrine Interactions, DOI 10.1007/978-3-319-41603-8_10

families. However, available treatments are far from ideal. Only a third of patients respond to the initial treatment, another third will get better only after several changes of medication and the rest will go on to be treatment resistant (Rapaport et al. 2003; Trivedi et al. 2006). Why is this? We still do not know why depression happens; neither do we clearly understand how antidepressants work. Much of the current understanding about the pathogenesis of major depression has come from animal models (Krishnan and Nestler 2011) as well as from peripheral (Felger et al. 2012) and central nervous system (CNS; Raison et al. 2010) circulating measurements from patients with depression. Due to the unique and complex features of human depression, the generation of valid and more insightful depression models has been less straightforward than modeling other disabling diseases (Krishnan and Nestler 2011). One possible approach is that of focusing on brain models, using neural cell lines. Undifferentiated or differentiated tumor-derived cells have been used as a translationally valid experimental model for several psychiatric disorders, including depression (Donnici et al. 2008; Alboni et al. 2013). However, such lines are limited in the cell types they can be made to resemble and may have major chromosomal abnormalities (Bray et al. 2012). Into this breach come new brain models of neural stem cells. Using multipotent fetal hippocampal progenitor cells (HPCs) and human induced pluripotent stem cells (iPSCs), we have mimicked clinically pertinent conditions to depressive disorders by combining depressogenic insults and antidepressant strategies (Anacker et al. 2011b, 2013a, b; Zunszain et al. 2012; Horowitz et al. 2015). Indeed, our outcomes provided evidence for the efficacy of such models in understanding the disorder as well as for giving more insights into antidepressants and their mechanisms of action. Particularly, we focused on neurogenesis as a potential candidate mechanism for the etiology of this condition as well as a substrate for antidepressant action.

The Neurogenesis Theory of Depression

A reduction in hippocampal neurogenesis, that is the birth of neurons from stem cells, has been suggested as one of the neurobiological alterations mediating the development of depressive-like behavior in animals, particularly under conditions of stress (David et al. 2009; Snyder et al. 2011; Surget et al. 2011). In the absence of effective neurogenesis, the depressive-like behavior elicited in animals by stress includes the hallmark abnormalities of clinical depression: increased hypothalamic-pituitary-adrenal (HPA) axis activity, glucocorticoid resistance (that is, impaired suppression of HPA axis activity by dexamethasone), anhedonia (assessed using the sucrose preference test), and behavioural despair (assessed using the forced swim test) (David et al. 2009; Snyder et al. 2011; Surget et al. 2011). Moreover, it has been suggested that an impaired neurogenesis may also precipitate depressive symptoms because of the lack of neurogenesis-dependent cognitive functions, such as the ability to enhance encoding of new memories and responding to contextual changes, which may be protective against behavioural despair in the face of repeated stressors (Sahay et al. 2011). Recent studies showing that the magnitude of adult neurogenesis in humans is probably larger than generally believed (Snyder and Cameron 2012; Spalding et al. 2013) provided even stronger support for the importance of neurogenesis and its proposed involvement in the association between stress and depression (Snyder et al. 2011).

Increased inflammation can also cause reductions in neurogenesis. Immune molecules, including interleukin-1beta (IL-1β), IL-6, interferon-alpha (IFN-α) and tumor necrosis factor-alpha (TNF-α) have been shown to be significantly upregulated in the peripheral blood of depressed patients (Howren et al. 2009; Dowlati et al. 2010). Particularly in the context of depression, IL-1β, IL6, IFN-α and IFN-γ have also been shown to easily move from the periphery into the brain (Dantzer et al. 2008; Najjar et al. 2013). Once they cross the brain-blood barrier, these molecules can alter distinct molecular and cellular mechanisms, including cell proliferation and neuronal maturation (Pickering and O'Connor 2007; Alboni et al. 2014) associated with complex cognitive processes, such as mood and learning functions (Makhija and Karunakaran 2013; Shigemoto-Mogami et al. 2014). In particular, using animal models, IL-1β, IL-18, IFN-α and TNF-α, have been shown to contribute to inhibition of synaptic plasticity and memory consolidation (Pickering and O'Connor 2007), causing similar impairments to those often reported in patients with major depressive disorder or in experimental models of depression (Pollak and Yirmiya 2002; Capuron and Miller 2004; Zunszain et al. 2013).

Furthermore, stress and inflammation interact. For example, in response to chronic IFN-α administration, patients with Hepatitis C Virus showed a hyper-reactivity of the HPA axis (Capuron et al. 2002; Raison et al. 2010). Most interestingly, around 30 % of those patients developed clinically significant depression (Raison et al. 2009), strengthening the notion that stress and inflammation might be among the crosstalk pathways leading to the pathogenesis of the depressive disorder. Indeed, it is of relevance in this context that stress and inflammation are among the different downstream molecular mechanisms that distinct antidepressants activate. Particularly, antidepressants from the selective serotonin reuptake inhibitor (SSRI) class, such as fluoxetine, have shown to normalize stress-induced HPA hyperactivity in rodents (Perera et al. 2011; Surget et al. 2011), whereas other serotonin-norepinephrine reuptake inhibitor (SNRI) antidepressants, such as venlafaxine, have normalized inflammatory alterations in cytokine-treated depressed patients (Capuron et al. 2002). However, irrespective of which distinct downstream molecular mechanisms specific antidepressants activate, those pathways may ultimately converge to stimulate neurogenesis, which is proposed as an essential substrate for antidepressant action (Schloesser et al. 2010).

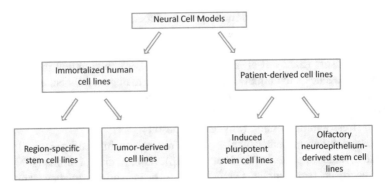

Fig. 1 The different range of neural cell models used to study depressive disorder and mechanisms of action of antidepressants

Experimental Approaches

Among the approaches available to investigate the complexity of depressive disorders, immortalized cell lines and patient-derived stem cell lines have proved to be relevant in vitro human neural cell models (Fig. 1). Immortalized cell lines allow molecular, developmental and pathophysiological mechanisms to be studied with considerable reliability. Examples of immortalized cell lines are region-specific neural stem cells and tumour-derived cells. An alternative cell-based approach is to derive and compare neural cells from patients and control individuals, particularly using iPSCs or olfactory neuroepithelium-derived cell lines. Indeed, the use of cells from patients allows a more attentive analysis of the pathological processes arising from the whole range of genetic susceptibility variants characterizing each individual (Srikanth and Young-Pearse 2014). Evidence has shown the significant incidence of distinct genetic polymorphisms in patients with psychiatric conditions, including depression (Cao et al. 2015; Chen et al. 2015; Wang et al. 2015), suggesting the importance of using such models to investigate genetic differences, which may allow for a predictive diagnosis of this disorder (Pasca et al. 2014). For the purpose of this chapter we will focus on immortalized stem cells and iPSC models. We will subsequently report our findings, providing evidence for their efficacy in understanding depressive disorders and antidepressant mechanisms of action.

Immortalized Human Neural Cell Lines

Neural Stem Cell Lines

Stem cells derived from human fetal brain are multipotent (i.e., they can give rise to a range of neurons and glia) and allow developmental and physiological processes to be studied more faithfully. Particularly, clonal neural stem cell lines can be generated by conditional immortalization, whereby a regulated gene that drives cell division is introduced into the cell's genome, allowing controlled expansion and differentiation (Pollock et al. 2006). Indeed, neural stem cell lines with normal chromosomes have been established from several human fetal brain regions, including cerebral cortex, hippocampus and striatum. This approach has several unique advantages. First, it delivers data from living human brain cells, not easily accessible in clinical samples; second, it can mimic a multitude of clinically relevant conditions within a tightly controlled experimental environment, providing a model system with which to explore the mechanisms of drug treatments for a variety of psychiatric disorders; and finally, it generates findings that are directly translatable in clinical samples (Bray et al. 2012). Using this approach, we have modeled "depression in a dish" using the cell line HPC0A07/03C (provided by ReNeuron Ltd, London), derived from the hippocampus, which allowed us to translate findings from bench-to-beside-and-back (Anacker et al. 2011b, 2013a, b; Zunszain et al. 2012; Horowitz et al. 2015). We will describe our observations in further detail.

Modeling the Role of Stress

As a first example of our translational approach, we managed to provide evidence for the detrimental role of stress on hippocampal neurogenesis. Impaired neurogenesis in rodents has recently been shown to contribute to the development of depressive-like behaviours, including anhedonia and behavioural despair in response to acute and chronic stressful insults (Zhu et al. 2014). Using our human hippocampal model, we showed that cortisol caused a reduction in the generation of new neurons via glucocorticoid receptor (GR)-dependent mechanisms, an effect which could be fully reverted by treatment with the SSRI sertraline. Indeed, subsequent stimulation with a GR-antagonist completely abolished the increase in neurogenesis induced by the antidepressant (Anacker et al. 2011b). Our model proved to be effective in providing further details from the complex interaction between stress and neuronal generation, proposing GR-dependent mechanisms as possible future targets of antidepressant drug treatment to overcome neurogenesis-related disturbances in depression (Anacker et al. 2011a).

Modeling the Role of Inflammation and Oxidative Stress

As a second example of the use of our model, we demonstrated the involvement of inflammation on hippocampal neurogenesis. Previous evidence had reported that in vitro stimulation with distinct cytokines, including IL-1β, IL-6, IFN-α and TNF-α, caused a significant alteration in both proliferation and neuronal maturation of human and animal cells (Borsini et al. 2015). Using our in vitro human neuronal model, we investigated the effect of two pro-inflammatory cytokines, IFN-α (our unpublished observation) and IL-1β (Zunszain et al. 2012), on neurogenesis. Findings showed that, upon treatment with both cytokines HPCs developed a "depressive phenotype" comprising reduced neurogenesis. Moreover, IL-1β was responsible for alterations in transcription pathways regulating the metabolism of tryptophan. Indeed, the inhibitory effects of IL-1β on neurogenesis were mediated, at least in part, by activation of the neurotoxic branch of the kynurenine pathway, one of the main pathways postulated to be involved in the development of depressive disorders (Baranyi et al. 2015).

A third example involved the use of tert-butylhydroxiperoxide (TBHP) to model oxidative stress. High levels of reactive oxygen species, shown in depressed patients, are known to affect cellular constituents, leading to neoepitopes and damage-associated molecular patterns that promote further immune responses (Bakunina et al. 2015). Cells treated with TBHP showed a dose-dependent increase in lipid peroxidation as well as reduced cell viability.

Studying Mechanism of Action of Antidepressants

Finally, we used this model to investigate the immunomodulatory properties of distinct compounds with antidepressant actions. We explored the effects of several conventional monoaminergic antidepressants and the omega-3 polyunsaturated fatty acids (n-3 PUFAs), eicosapentanoic acid (EPA) and docosahexanoic acid (DHA), on HPCs treated with the inflammatory and depressogenic IL-1β. In contrast to sertraline and DHA, which had pro-inflammatory properties, venlafaxine and EPA were shown to have anti-inflammatory effects via decreasing distinct cytokines, including IL-6, IL-8 and IP-10 (Horowitz et al. 2015). In addition, these compounds showed differential effects on neurogenesis. Again, the findings demonstrate the efficacy of this model for studying specific mechanisms of action of drugs with antidepressant action.

Tumour-Derived Cell Lines

These lines, with an ability to expand quite readily in culture, provide a standardized and potentially limitless alternative to study intracellular mechanisms of antidepressant action. Currently, the most commonly used human neural cell line is SH-SY5Y. This line displays neuronal properties, including neurite outgrowth, neurotransmitter synthesis and relevant receptor expression. The SH-SY5Y line has been widely used to study intracellular mechanisms of different antidepressant action, including the SSRI sertraline, the selective norepinephrine reuptake inhibitor (SNRI) desipramine and the norepinephrine–dopamine reuptake inhibitor (NDRI) bupropion (Lin 2015).

Patient-Derived Neural Cells

IPSCs

Among the patient-derived cell models, iPSC technology provides distinct cell types that are considered to be central to psychiatric disorders, such as those of the cortex and the hippocampus (Jaworska et al. 2015). Indeed, primary somatic cells, typically from skin, can be taken from an individual and reprogrammed into pluripotent stem cells that can give rise to all of the cell types that characterize the body, including those of the CNS. By capturing a patient's entire genome and any possible epigenetic variations, iPSCs constitute a unique source of material for studying neurodevelopmental features of psychiatric disorders in vitro.

Reports are now beginning to emerge in which this technology has been applied to cells taken from psychiatric patients. For example, human keratinocytes from healthy controls and patients with bipolar depression have been reprogrammed into cortical neurons. When compared with control cells, neurons derived from patients with bipolar depression showed an alteration in the expression of transcripts that regulate Hedgehog signaling (Cheung et al. 2009), as well as modulations in key components of the mTOR pathway (O'Shea and McInnis 2015), which have both been shown to be among the mechanisms involved in the development of depressive disorders (Rajendran et al. 2009; Ignacio et al. 2015). Using iPSC-derived forebrain neurons, we showed that ketamine, known to have fast-acting antidepressant efficacy in treatment-resistant patients, was able to rescue the detrimental effects produced by treatment with IL-1β and to increase the number of presynaptic and postsynaptic proteins.

Olfactory Neuroepithelium-Derived Stem Cells

Cells from the olfactory mucosa, which can be extracted through biopsy, can easily propagate, forming neurospheres of neural stem cells and differentiating neural progenitor cells. Although there are no studies using olfactory neuroepithelium-derived cells from depressed patients, they have been used to study schizophrenia (Matigian et al. 2010; Fan et al. 2012), suggesting the importance of using such models to investigate genetic differences, which may allow for a predictive diagnosis of depression in certain individuals.

Conclusions and Limitations of the Cell Models

Stem cell-based approaches to study psychiatric disorders are advancing on two main fronts. On one hand, clonal cell lines which accurately model the CNS are being used in controlled experiments to assess the mechanisms of antidepressant action for psychiatric disorders, which might in the short term lead to advancements in therapeutic strategies for these conditions. On the other hand, patient-derived cells and cells from control patients allow the study of pathological processes deriving from the multiple genetic susceptibility variants, which can now be investigated with more accuracy. However, both cell models have some limitations that need to be pointed out. Although human neural cell lines can be used to investigate the molecular and cellular functions of individual susceptibility genes, they do not capture the many genetic variables that contribute to the development of psychiatric disorders (Bray et al. 2012). Patient-derived cell lines offer the advantage of capturing each individual's whole genome, but there is limited knowledge as to which cell types are most relevant to study specific psychiatric conditions (Sandoe and Eggan 2013). In addition, while iPSC technology can model the effects of medications, these cells may lose the effects of environmental influences that may contribute to the development of the psychiatric illnesses, such as stressors or negative life events (Okano and Yamanaka 2014).

Although cell-based models are still unable to elucidate the molecular complexity of psychiatric illnesses, the enormous progress in stem cell technologies has revolutionized the field of "in vitro disease modeling," providing not only a window into the mechanisms underlying the depressive disorder but also a platform for screening novel therapeutic strategies for the prevention and treatment of this condition.

References

Alboni S, Gibellini L, Montanari C, Benatti C, Benatti S, Tascedda F, Brunello N, Cossarizza A, Pariante CM (2013) N-acetyl-cysteine prevents toxic oxidative effects induced by IFN-alpha in human neurons. Int J Neuropsychopharmacol 16:1849–1865

Alboni S, Montanari C, Benatti C, Sanchez-Alavez M, Rigillo G, Blom JM, Brunello N, Conti B, Pariante MC, Tascedda F (2014) Interleukin 18 activates MAPKs and STAT3 but not NF-kappaB in hippocampal HT-22 cells. Brain Behav Immun 40:85–94

Anacker C, Zunszain PA, Carvalho LA, Pariante CM (2011a) The glucocorticoid receptor: pivot of depression and of antidepressant treatment? Psychoneuroendocrinology 36:415–425

Anacker C, Zunszain PA, Cattaneo A, Carvalho LA, Garabedian MJ, Thuret S, Price J, Pariante CM (2011b) Antidepressants increase human hippocampal neurogenesis by activating the glucocorticoid receptor. Mol Psychiatry 16:738–750

Anacker C, Cattaneo A, Luoni A, Musaelyan K, Zunszain PA, Milanesi E, Rybka J, Berry A, Cirulli F, Thuret S, Price J, Riva MA, Gennarelli M, Pariante CM (2013a) Glucocorticoid-related molecular signaling pathways regulating hippocampal neurogenesis. Neuropsychopharmacology 38:872–883

Anacker C, Cattaneo A, Musaelyan K, Zunszain PA, Horowitz M, Molteni R, Luoni A, Calabrese F, Tansey K, Gennarelli M, Thuret S, Price J, Uher R, Riva MA, Pariante CM (2013b) Role for the kinase SGK1 in stress, depression, and glucocorticoid effects on hippocampal neurogenesis. Proc Natl Acad Sci USA 110:8708–8713

Bakunina N, Pariante CM, Zunszain PA (2015) Immune mechanisms linked to depression via oxidative stress and neuroprogression. Immunology. doi:10.1111/imm.12443

Baranyi A, Meinitzer A, Breitenecker RJ, Amouzadeh-Ghadikolai O, Stauber R, Rothenhausler HB (2015) Quinolinic acid responses during interferon-alpha-induced depressive symptomatology in patients with chronic hepatitis C infection—a novel aspect for depression and inflammatory hypothesis. PLoS One 10:e137022

Borsini A, Zunszain PA, Thuret S, Pariante CM (2015) The role of inflammatory cytokines as key modulators of neurogenesis. Trends Neurosci 38:145–157

Bray NJ, Kapur S, Price J (2012) Investigating schizophrenia in a "dish": possibilities, potential and limitations. World Psychiatry 11:153–155

Cao S, Li H, Lou L, Xie Z, Zhao X, Pang J, Sui J, Xie G (2015) Association study between 5-HT2A and NET gene polymorphisms and recurrent major depression disorder in Chinese Han population. Pak J Pharm Sci 28:1101–1108

Capuron L, Miller AH (2004) Cytokines and psychopathology: lessons from interferon-alpha. Biol Psychiatry 56:819–824

Capuron L, Hauser P, Hinze-Selch D, Miller AH, Neveu PJ (2002) Treatment of cytokine-induced depression. Brain Behav Immun 16:575–580

Chen J, Wang M, Waheed Khan RA, He K, Wang Q, Li Z, Shen J, Song Z, Li W, Wen Z, Jiang Y, Xu Y, Shi Y, Ji W (2015) The GSK3B gene confers risk for both major depressive disorder and schizophrenia in the Han Chinese population. J Affect Disord 185:149–155

Cheung HO, Zhang X, Ribeiro A, Mo R, Makino S, Puviindran V, Law KK, Briscoe J, Hui CC (2009) The kinesin protein Kif7 is a critical regulator of Gli transcription factors in mammalian hedgehog signaling. Sci Signal 2:ra29

Dantzer R, O'Connor JC, Freund GG, Johnson RW, Kelley KW (2008) From inflammation to sickness and depression: when the immune system subjugates the brain. Nat Neurosci 9:46–56

David DJ, Samuels BA, Rainer Q, Wang JW, Marsteller D, Mendez I, Drew M, Craig DA, Guiard BP, Guilloux JP, Artymyshyn RP, Gardier AM, Gerald C, Antonijevic IA, Leonardo ED, Hen R (2009) Neurogenesis-dependent and -independent effects of fluoxetine in an animal model of anxiety/depression. Neuron 62:479–493

Donnici L, Tiraboschi E, Tardito D, Musazzi L, Racagni G, Popoli M (2008) Time-dependent biphasic modulation of human BDNF by antidepressants in neuroblastoma cells. BMC Neurosci 9:61

Dowlati Y, Herrmann N, Swardfager W, Liu H, Sham L, Reim EK, Lanctot KL (2010) A meta-analysis of cytokines in major depression. Biol Psychiatry 67:446–457

Fan Y, Abrahamsen G, McGrath JJ, Mackay-Sim A (2012) Altered cell cycle dynamics in schizophrenia. Biol Psychiatry 71:129–135

Felger JC, Cole SW, Pace TW, Hu F, Woolwine BJ, Doho GH, Raison CL, Miller AH (2012) Molecular signatures of peripheral blood mononuclear cells during chronic interferon-alpha treatment: relationship with depression and fatigue. Psychol Med 42:1591–1603

Horowitz MA, Wertz J, Zhu D, Cattaneo A, Musaelyan K, Nikkheslat N, Thuret S, Pariante CM, Zunszain PA (2015) Antidepressant compounds can be both pro- and anti-inflammatory in human hippocampal cells. Int J Neuropsychopharmacol 18(3). doi:10.1093/ijnp/pyu076

Howren MB, Lamkin DM, Suls J (2009) Associations of depression with C-reactive protein, IL-1, and IL-6: a meta-analysis. Psychosom Med 71:171–186

Ignacio ZM, Reus GZ, Arent CO, Abelaira HM, Pitcher MR, Quevedo J (2015) New perspectives on the involvement of mTOR in depression as well as in the action of antidepressant drugs. Brit J Clin Phamacol. doi:10.1111/bcp.12845

Jaworska N, Yucel K, Courtright A, MacMaster FP, Sembo M, MacQueen G (2015) Subgenual anterior cingulate cortex and hippocampal volumes in depressed youth: the role of comorbidity and age. J Affect Disord 190:726–732

Krishnan V, Nestler EJ (2011) Animal models of depression: molecular perspectives. Curr Top Behav Neurosci 7:121–147

Lin PY (2015) Regulation of proteolytic cleavage of brain-derived neurotrophic factor precursor by antidepressants in human neuroblastoma cells. Neuropsychiatr Dis Treat 11:2529–2532

Makhija K, Karunakaran S (2013) The role of inflammatory cytokines on the aetiopathogenesis of depression. Aust N Z J Psychiatry 47:828–839

Matigian N, Matigian N, Abrahamsen G, Sutharsan R, Cook AL, Vitale AM, Nouwens A, Bellette B, An J, Anderson M, Beckhouse AG, Bennebroek M, Cecil R, Chalk AM, Cochrane J, Fan Y, Féron F, McCurdy R, McGrath JJ, Murrell W, Perry C, Raju J, Ravishankar S, Silburn PA, Sutherland GT, Mahler S, Mellick GD, Wood SA, Sue CM, Wells CA, Mackay-Sim A (2010) Disease-specific, neurosphere-derived cells as models for brain disorders. Dis Model Mech 3:785–798

Najjar S, Pearlman DM, Devinsky O, Najjar A, Zagzag D (2013) Neurovascular unit dysfunction with blood-brain barrier hyperpermeability contributes to major depressive disorder: a review of clinical and experimental evidence. J Neuroinflammation 10:142

Okano H, Yamanaka S (2014) iPS cell technologies: significance and applications to CNS regeneration and disease. Mol Brain 7:22. doi:10.1186/1756-6606-7-22

O'Shea KS, McInnis MG (2015) Induced pluripotent stem cell (iPSC) models of bipolar disorder. Neuropsychopharmacology 40:248–249

Pasca SP, Panagiotakos G, Dolmetsch RE (2014) Generating human neurons in vitro and using them to understand neuropsychiatric disease. Ann Rev Neurosci 37:479–501

Perera TD, Dwork AJ, Keegan KA, Thirumangalakudi L, Lipira CM, Joyce N, Lange C, Higley JD, Rosoklija G, Hen R, Sackeim HA, Coplan JD (2011) Necessity of hippocampal neurogenesis for the therapeutic action of antidepressants in adult nonhuman primates. PLoS One 6:e17600

Pickering M, O'Connor JJ (2007) Pro-inflammatory cytokines and their effects in the dentate gyrus. Prog Brain Res 163:339–354

Pollak Y, Yirmiya R (2002) Cytokine-induced changes in mood and behaviour: implications for 'depression due to a general medical condition', immunotherapy and antidepressive treatment. Int J Neuropsychopharmacol 5:389–399

Pollock K, Stroemer P, Patel S, Stevanato L, Hope A, Miljan E, Dong Z, Hodges H, Price J, Sinden JD (2006) A conditionally immortal clonal stem cell line from human cortical neuroepithelium for the treatment of ischemic stroke. Exp Neurol 199:143–155

Raison CL, Borisov AS, Majer M, Drake DF, Pagnoni G, Woolwine BJ, Vogt GJ, Massung B, Miller AH (2009) Activation of central nervous system inflammatory pathways by interferon-alpha: relationship to monoamines and depression. Biol Psychiatry 65:296–303

Raison CL, Borisov AS, Woolwine BJ, Massung B, Vogt G, Miller AH (2010) Interferon-alpha effects on diurnal hypothalamic-pituitary-adrenal axis activity: relationship with proinflammatory cytokines and behavior. Mol Psychiatry 15:535–547

Rajendran R, Jha S, Fernandes KA, Banerjee SB, Mohammad F, Dias BG, Vaidya VA (2009) Monoaminergic regulation of Sonic hedgehog signaling cascade expression in the adult rat hippocampus. Neurosci Lett 453:190–194

Rapaport MH, Schneider LS, Dunner DL, Davies JT, Pitts CD (2003) Efficacy of controlled-release paroxetine in the treatment of late-life depression. J Clin Psychiatry 64:1065–1074

Sahay A, Scobie KN, Hill AS, O'Carroll CM, Kheirbek MA, Burghardt NS, Fenton AA, Dranovsky A, Hen R (2011) Increasing adult hippocampal neurogenesis is sufficient to improve pattern separation. Nature 472:466–470

Sandoe J, Eggan K (2013) Opportunities and challenges of pluripotent stem cell neurodegenerative disease models. Nat Neurosci 16:780–789

Schloesser RJ, Lehmann M, Martinowich K, Manji HK, Herkenham M (2010) Environmental enrichment requires adult neurogenesis to facilitate the recovery from psychosocial stress. Mol Psychiatry 15:1152–1163

Shigemoto-Mogami Y, Hoshikawa K, Goldman JE, Sekino Y, Sato K (2014) Microglia enhance neurogenesis and oligodendrogenesis in the early postnatal subventricular zone. J Neurosci 34:2231–2243

Snyder JS, Cameron HA (2012) Could adult hippocampal neurogenesis be relevant for human behavior? Behav Brain Res 227:384–390

Snyder JS, Soumier A, Brewer M, Pickel J, Cameron HA (2011) Adult hippocampal neurogenesis buffers stress responses and depressive behaviour. Nature 476:458–461

Spalding KL, Bergmann O, Alkass K, Bernard S, Salehpour M, Huttner HB, Bostrom E, Westerlund I, Vial C, Buchholz BA, Possnert G, Mash DC, Druid H, Frisen J (2013) Dynamics of hippocampal neurogenesis in adult humans. Cell 153:1219–1227

Srikanth P, Young-Pearse TL (2014) Stem cells on the brain: modeling neurodevelopmental and neurodegenerative diseases using human induced pluripotent stem cells. J Neurogenet 28:5–29

Surget A, Tanti A, Leonardo ED, Laugeray A, Rainer Q, Touma C, Palme R, Griebel G, Ibarguen-Vargas Y, Hen R, Belzung C (2011) Antidepressants recruit new neurons to improve stress response regulation. Mol Psychiatry 16:1177–1188

Trivedi MH, Rush AJ, Wisniewski SR, Nierenberg AA, Warden D, Ritz L, Norquist G, Howland RH, Lebowitz B, McGrath PJ, Shores-Wilson K, Biggs MM, Balasubramani GK, Fava M, Team SS (2006) Evaluation of outcomes with citalopram for depression using measurement-based care in STAR*D: Implications for clinical practice. Am J Psychiatry 163:28–40

Wang Y, Sun N, Li S, Du Q, Xu Y, Liu Z, Zhang K (2015) A genetic susceptibility mechanism for major depression: combinations of polymorphisms defined the risk of major depression and subpopulations. Medicine (Baltimore) 94:e778

Zhu S, Wang J, Zhang Y, Li V, Kong J, He J, Li XM (2014) Unpredictable chronic mild stress induces anxiety and depression-like behaviors and inactivates AMP-activated protein kinase in mice. Brain Res 1576:81–90

Zunszain PA, Anacker C, Cattaneo A, Choudhury S, Musaelyan K, Myint AM, Thuret S, Price J, Pariante CM (2012) Interleukin-1beta: a new regulator of the kynurenine pathway affecting human hippocampal neurogenesis. Neuropsychopharmacology 37:939–949

Zunszain PA, Hepgul N, Pariante CM (2013) Inflammation and depression. Curr Top Behav Neurosci 14:135–151

Perspective on Stem Cells in Developmental Biology, with Special Reference to Neuroendocrine Systems

Karine Rizzoti, Carlotta Pires, and Robin Lovell-Badge

Abstract In the embryo, organs gradually take shape as tissue progenitors, proliferate, differentiate and are organised, via cellular interactions, in tri-dimensional functional structures. Most adult organs retain a cell population sharing important similarities with embryonic progenitors, which includes the ability to both self-renew and to differentiate into the full range of the specialised cell types corresponding to the organ in which they reside. These two essential properties define them as adult stem cells (AdSC). Their characterization in different contexts provides a better understanding of cell turnover modalities for organ function, which is important for tumorigenesis, because adult tissue stem cells can give rise to cancer stem cells. However, it is also relevant to regenerative medicine, because stem cells can be transplanted or manipulated in vivo to restore missing cells. The successful reprogramming of somatic cells into induced pluripotent stem cells (iPSC) (Takahashi and Yamanaka, Cell 126:663–676, 2006), resembling pluripotent embryonic stem cells (ESC), opened alternative strategies for cell replacement. The required cell type can be differentiated in vitro from expandable progenitors and transplanted where required. This approach is particularly promising, because genetic defects can potentially be repaired in vitro prior to their engraftment; moreover, a large number of cells can be produced (Fox et al., Science 345:1247391, 2014). However, the transplanted cells have to be able to functionally integrate into the deficient organ. To potentially alleviate this problem, tri-dimensional culture systems have been recently developed where mini-organs, or organoids, can be obtained in vitro. These also represent unique models for disease modelling and drug screening (Huch and Koo, Development 142:3113–3125, 2015). Within neuroendocrine systems, the hypothalamo-pituitary axis has a crucial role for body homeostasis. It also regulates functions such as growth, reproduction, stress and, more generally, metabolism. The hypothalamus centralizes information from the periphery and other brain regions to regulate pituitary hormone secretions and to control appetite, sleep and aging. Its functions

K. Rizzoti (✉) • R. Lovell-Badge (✉)
The Francis Crick Institute, Mill Hill Laboratory, The Ridgeway, London NW7 1AA, UK
e-mail: karine.rizzoti@crick.ac.uk; robin.lovellbadge@crick.ac.uk

C. Pires
University of Copenhagen, Gronnegaardsvej 7, Frederiksberg 1870, Denmark

© The Author(s) 2016
D. Pfaff, Y. Christen (eds.), *Stem Cells in Neuroendocrinology*, Research and Perspectives in Endocrine Interactions, DOI 10.1007/978-3-319-41603-8_11

are exerted through secretion of neurohormones and manipulation of these could be of interest not only for the treatment of obesity and sleep disorders but also to alleviate aging-related conditions. Pituitary hormone deficiencies can originate from defects affecting the hypothalamus, pituitary, or both. Today these deficiencies are treated by substitution or replacement therapies, which are costly and have side effects. AdSC have recently been characterized in the hypothalamus and pituitary (Castinetti et al., Endocr Rev 32:453–471, 2011; Bolborea and Dale, Trends Neurosci 36:91–100, 2013). Moreover, hypothalamic neurons and pituitary endocrine cells have been differentiated in vitro from ESC and iPSC and successfully transplanted. These promising routes for development of regenerative medicine to restore or manipulate the function of the hypothalamo-pituitary axis will be discussed here, and further challenges considered.

Introduction

It is now clear that multipotency, once thought to be largely restricted to embryonic stages and to a few stem cell types in the adult, is preserved postnatally, as most adult tissues comprise a population of undifferentiated cells. These are characterized by their ability to undergo asymmetric divisions, underlying self-renewal and an ability to have differentiated cell progeny, the two essential properties of stem cells. Pluripotency, the ability to produce all adult cell types, can be "recovered" in vitro by reprogramming somatic cells into induced pluripotent stem cells (iPSC; Takahashi and Yamanaka 2006). In addition, transdifferentiation to another cell type or to multipotency can be forced onto differentiated cell types by activating expression of specific factors or via cell fate conversion with chemicals and/or growth factors (Vierbuchen and Wernig 2012). While these discoveries are by themselves challenging our view of cellular plasticity and its limits, for medicine they mean that therapies can progress from or incorporate both a pharmacological approach and a cellular one, where abnormal, deficient or entirely missing cells can be replaced (Fox et al. 2014). Cellular reprogramming implies, in particular, the possibility of autologous transplantation, alleviating at least some aspects of immune rejection that, in conjunction with genome editing tools (Hsu et al. 2014), hold great promise for deriving, repairing, differentiating and transplanting back functional, patient-specific cells. In addition, the derivation of patient-specific cells by itself represents a great advance toward in vitro disease modelling and drug screening (van de Wetering et al. 2015). In the clinic, some promising results have already been obtained in pilot studies using retinal pigment cells derived from human ESC (Schwartz et al. 2015), but other trials have also highlighted the many challenges to be overcome (Steinbeck and Studer 2015). It is likely that further development of three-dimensional (3D) organoids (Huch and Koo 2015) and tissue engineering, combining several different cell types for formation of vascularized mini-organs, will greatly enhance transplantation efficiency and functional repair (Shamir and Ewald 2014). Finally, another important aspect to consider for use of these therapies is the financial one. Numerous qualitative criteria have to be satisfied to allow clinical use of stem cells. Currently, and for good reasons, these criteria have to be stringent (French et al. 2015) and therefore expensive. Although experience, scaling up, and automation might lead to a relaxation of requirements and to reduced costs, we still need to question today how feasible it will be to routinely offer such treatments in the future. It would be unfair if treatments were only available to the very privileged few.

We will focus our attention here on stem cells of the hypothalamo-pituitary axis, and why and how these could be used for regenerative medicine. The hypothalamo-pituitary axis, while a crucial regulator of homeostasis, also ensures that the organism responds appropriately to changing physiological situations such as puberty, pregnancy, lactation, and stress. While many hypothalamic roles are executed through the pituitary, with the seven different hormones secreted by the gland affecting most physiological processes, some hypothalamic neurons also directly interact with other regions of the brain to control appetite (Schneeberger et al. 2014), sleep and wakefulness (Konadhode et al. 2014). Finally, it has recently been proposed that systemic aging is initiated in the hypothalamus (Zhang et al. 2013). Consequently, pathological situations leading to hypothalamic and/or pituitary hormone imbalance have pleiotropic consequences, resulting in significant morbidity and even mortality. These can be congenital or acquired and originate from a defective pituitary, hypothalamus or both; moreover, in most cases they are of unknown aetiology (Kelberman et al. 2009). The most frequent congenital defect is isolated growth hormone deficiency, with a prevalence of 1 in 3500–10,000 births (Kelberman et al. 2009). Later in life, acquired deficits develop mainly as a consequence of pituitary tumors, mostly affecting lactotrophs, or from brain damage (Hannon et al. 2013). Hormone substitution or replacement therapies are available to treat pituitary deficiencies. However, these are not optimal, first because they do not reproduce normal physiological secretion patterns, secondly because they are associated with side effects (Alatzoglou et al. 2014), and thirdly, they are not inexpensive; hGH treatment in the UK costs up to £12,000/year for children in whom treatments can last for several years. Being able to provide secreting endocrine cells would, therefore, represent a significant advance, particularly for pediatric patients where the pituitary is primarily affected but perhaps also for those where an hypothalamic deficiency is thought to be responsible, such as Prader-Willi syndrome, or more generally where a hypothalamic defect leads to formation of an underdeveloped pituitary gland (Rizzoti et al. 2004). Pituitary hormones can, in turn, also feed back on the formation of hypothalamic neuronal circuitry (Sadagurski et al. 2015); therefore, they may be able to improve some hypothalamic defects. Manipulating or restoring hypothalamic function would be beneficial in cases of infertility and obesity, which has already been shown using cell or tissue transplants (see below), but also for sleep disorders; importantly, it might lighten some age-associated conditions.

In this perspective we will first briefly describe the morphogenesis of the ventral diencephalon, from which the hypothalamus derives, and of Rathke's pouch (RP), the pituitary anlagen. The transition between tissue progenitor and stem cell phenotype will then be discussed in the context of the developing pituitary. Recently, populations of AdSCs have been described and partially characterized in both compartments of the axis (Lee et al. 2012; Li et al. 2012; Andoniadou et al. 2013; Haan et al. 2013; Rizzoti et al. 2013; Robins et al. 2013a, b; Castinetti et al. 2011) and their characterization will be presented here, with reference to other chapters of this book. Finally, exciting progress has been made toward the use of stem cells for regenerative medicine in the axis: both hypothalamic neurons (Wataya et al. 2008; Merkle et al. 2015; Wang et al. 2015) and pituitary endocrine cells (Suga et al. 2011; Dincer et al. 2013) have been obtained in vitro from stem cells and successfully transplanted in mice. We will discuss these reports and the future challenges toward clinical use of stem cells in the hypothalamo-pituitary axis.

Morphogenesis of the Hypothalamo-Pituitary Axis (Fig. 1)

Hypothalamus

In the embryo, the hypothalamus develops from the ventral diencephalon. During the specification of the neural plate, at 8 days post-coitum (dpc) in mice, the prospective ventral diencephalon is situated at the midline, in the rostral-most position. It is in contact with the future pituitary, which is present as the hypophyseal placode at this stage, in the adjacent ectoderm at the anterior neural ridge (see below). As the neural plate bends to close (McShane et al. 2015), increased proliferation of the telencephalic progenitors versus those in the ventral diencephalon causes an apparent posterior shift of the prospective hypothalamus, located ventrally and posterior to the telencephalic vesicles in 9.5-dpc mouse embryos. A localized evagination of the ventral diencephalon toward the underlying developing pituitary becomes apparent at 9.5 dpc. From this particular region, called the infundibulum, the median eminence (ME), pituitary stalk and pituitary posterior lobe will develop. The ME, located at the floor of the third ventricle, is a circumventricular organ, which means that the blood-brain barrier is interrupted at its level. Hypothalamic neuro-hormones collect here into a bed of fenestred capillaries belonging to the hypohyseal portal system, which then transport them to the pituitary. Within the ME, special glial cells, the tanycytes, regulate neurohormone secretion at the axon termini levels (Prevot et al. 2010). These also integrate peripheral information by, for example, sensing glucose concentration and responding to thyroid hormone (Bolborea and Dale 2013; Ebling 2015). Moreover, and importantly, at least a proportion of tanycytes have been shown to comprise hypothalamic stem cells (see below). These cells therefore perform crucial regulatory roles at this blood-brain interface in the short term, by regulating neurohormone secretion, but also in the long term, by modulating hypothalamic cell numbers. Some hypothalamic

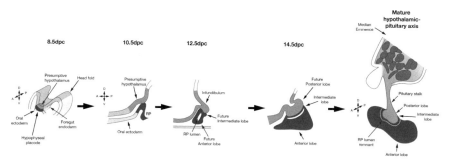

Fig. 1 Development and function of the murine pituitary-hypothalamic axis. Rostral to the presumptive hypothalamus, pituitary development is initiated with the appearance of the hypophyseal placode at 8.5 dpc in the mouse. At 9.5 dpc, the placode invaginates to become the Rathke's pouch (RP). Within the diencephalon, the infundibulum evaginates towards RP by 10.5 dpc. It will give rise to the median eminence (ME), the pituitary stalk and the posterior lobe, whereas the anterior and intermediate lobes originate from RP. Gradually, hypothalamic neurons forming the different nuclei differentiate. Postnatally, the hypothalamus centralizes peripheral information and controls, through release of hypophysiotropic hormones, pituitary endocrine secretions. Hypothalamic peptide hormones can reach the gland directly, such as oxytocin and vasopressin secreted directly in the posterior lobe, or via the hypophyseal portal system. Hypothalamic neuropeptides (GnRH, gonadotrophin releasing hormone, *GHRH* GH releasing hormone, *TRH* TSH releasing hormone, *CRH* corticotropin releasing hormone and the inhibitory SST, somatostatin) are secreted at the ME and collected by capillaries. Anterior pituitary endocrine cells comprise lactotrophs (producing prolactin, Prl), gonadotrophs (producing luteinizing hormone, LH, and follicle stimulating hormone, FSH), thyrotrophs (producing thyroid stimulating hormone, TSH), corticotrophs (producing adenocorticotropic hormone, ACTH, proteolytically cleaved from proopiomelanocortin, POMC) and somatotrophs (producing growth hormone, GH). Adapted from Rizzoti (2015)

neurons, namely those secreting oxytocin and vasopressin, reach directly into the posterior lobe of the pituitary; their axons, along with the portal system capillaries, are located within the pituitary stalk. Pituicytes, glial cells present both in the stalk and posterior pituitary, are proposed to regulate neurosecretion at these levels. Tanycytes and pituicytes both originate from the infundibulum, and no neurons are produced in this domain (Goto et al. 2015). The specification of the infundibulum in the embryo is complex and relies on an antagonism between members of the bone morphogenetic protein (BMP) family, present in the future infundibulum, and the secreted molecule Sonic Hedgehog (SHH), which is excluded from it (Zhao et al. 2012; Trowe et al. 2013). Members of the fibroblast growth factor (FGF) family are in turn required for infundibular cell expansion (Pearson et al. 2011). The transcription factor LHX2 is necessary for formation of the infundibulum in the embryo and later on for expression of the RAX transcription factor, which is essential for ventro-medial hypothalamic development (Lu et al. 2013) and the emergence of tanycytes (Salvatierra et al. 2014). The NOTCH pathway is also required for correct morphogenesis of this region (Goto et al. 2015), but it is unclear at present how these different signalling pathways are integrated.

As neurogenesis takes place, hypothalamic neurons differentiate and migrate to form groups of neurons, or nuclei. Nuclei located the furthest away from the third ventricle are formed first, whereas the latest to differentiate are closest. The GnRH neurons are the only exception, as these originate from the olfactory placode and migrate through the developing brain to reach their destination around the ME, where they control reproductive function (Stevenson et al. 2013) and aging (Zhang et al. 2013). Each nucleus regulates different physiological functions, such as circadian rhythms by the supra-chiasmatic nucleus or feeding behaviour by the arcuate. Hypothalamic morphogenesis is complex because of its organization, which lacks clear morphological landmarks; this explains why we know comparatively less about it compared to other regions of the brain where the anatomy is somehow simpler. Mechanisms of hypothalamic neurogenesis are beyond the scope of this perspective and have been recently comprehensively reviewed (Bedont et al. 2015; Pearson and Placzek 2013).

Pituitary

The first sign of pituitary development is the appearance at 8 dpc of an ectodermal thickening in the rostral-most position, in the anterior neural ridge just in front of the future ventral diencephalon. This domain, the hypophyseal placode, forms along with the olfactory, optic and otic placodes (Soukup et al. 2013; Schlosser et al. 2014). As the presumptive diencephalon is apparently shifted posteriorly at 9.5 dpc, the hypophyseal placode is present just underneath and has become the pituitary anlagen or RP. RP is an epithelial invagination that forms in continuity with the oral ectoderm and extends dorsally toward the ventral diencephalon. RP is also in contact with the anterior border of foregut endoderm (Khonsari et al. 2013), and both dorsally and ventrally secreted signals are crucial for proper RP morphogenesis. SHH, independently of its role for formation of the infundibulum, is required for RP development, but it is not expressed in the pouch (Wang et al. 2010). The infundibulum is crucial for induction and maintenance of the pouch at these early stages, predominantly through secretion of BMP and FGF signals, respectively (Ericson et al. 1998; Treier et al. 1998). RP grows rapidly and by 11.5 dpc it also separates from the oral ectoderm. The three pituitary lobes do not have the same embryonic origin because anterior and intermediate lobes are RP derivatives, whereas the posterior one has a neurectodermal origin.

As development progresses, the main wave of cell cycle exit in the developing pituitary, correlating with endocrine cell commitment and differentiation, occurs between 11.5 and 13.5 dpc (Japon et al. 1994; Davis et al. 2011). As cells exit the cell cycle, they adopt more ventral and lateral locations and loose epithelial characteristics, reminiscent of epithelial to mesenchymal transition (Himes and Raetzman 2009). Endocrine cells then gradually organize in homotypic endocrine networks, resulting in more efficient and coordinated hormonal secretion in the adult (Mollard et al. 2012). Pituitary cell fate specification is also discussed in Drouin (2016).

In rodents, the gland will undergo a phase of important growth during the first weeks of life. This is characterized by high levels of proliferation, both in progenitors and in endocrine cells. Endocrine cells retain the capability to proliferate throughout life, but they do so rarely (Levy 2002). After birth, the hypothalamus will take control of pituitary endocrine maturation, secretions, and perhaps also stem cell homeostasis, but it is not clear exactly when and how its influence starts to be exerted.

Transition Between Tissue Progenitor and Stem Cell Phenotype in the Developing Pituitary

Early RP progenitors express the transcription factor SOX2 (Fauquier et al. 2008), and lineage-tracing experiments using a $Sox2^{creERT2}$ allele have shown that SOX2-positive progenitors give rise to all endocrine cell types in the pituitary (Andoniadou et al. 2013; Rizzoti et al. 2013). In addition, embryonic progenitors give rise to adult pituitary stem cells (Rizzoti et al. 2013). From 14.5 dpc and towards the end of gestation in the mouse, SOX9 is upregulated in the pituitary, exclusively in SOX2-positive cells. Both proteins will remain co-expressed postnatally in pituitary stem cells (Fauquier et al. 2008; Rizzoti et al. 2013). While SOX2 and SOX9 belong to the same group of transcriptional regulators, sequence homology assigns them to different sub-families (SOXB1 and SOXE, respectively); they therefore have different partners, target genes and hence, roles (Kamachi and Kondoh 2013). SOX9 is required for maintenance of different ectodermal stem cell populations, such as hair, retinal, neural crest, and neural stem cells (NSC) (Sarkar and Hochedlinger 2013). In the developing neuroepithelium, SOX9 is both required and sufficient for formation of multipotent NSC, seemingly given them the ability to give rise to glial cell types and not just neurons (Scott et al. 2010). In the adult, the protein is involved in the maintenance of NSC and their multipotentiality (Scott et al. 2010).

In the developing pituitary, we observe that up-regulation of SOX9 correlates with a decreased proliferative potential in pituitary progenitors, with the percentage of SOX2-positive proliferating cells reduced by a factor of two between 12.5 and 16.5 dpc (Fig. 2). In addition, lineage-tracing experiments, using either a $Sox9^{ires-creERT2}$ allele or the $Sox2^{creERT2}$ allele, but where it is induced coincident with the upregulation of SOX9 in pituitary progenitors, show that the SOX9;SOX2-positive progenitors also tend to differentiate less often (Rizzoti et al. 2013). Therefore, the expression of SOX9 correlates both with reduced levels of proliferation and differentiation, two characteristics of pituitary stem cells that distinguish them from embryonic progenitors. So far, deletion of the gene during pituitary development has not revealed a significant role for SOX9 in the gland before birth (unpublished data); therefore, its function in pituitary progenitors does not parallel what is observed in the developing CNS. Further investigations are now required to examine its function postnatally. Analysis of its target genes may be informative to characterise the transition between tissue progenitor and stem cell phenotypes.

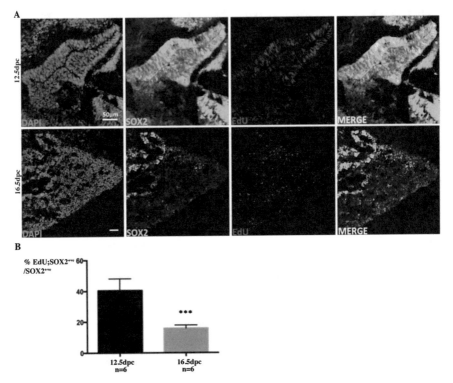

Fig. 2 Rate of proliferation of pituitary embryonic progenitors before and after SOX9 up-regulation. (**a**) Assessment of proliferative rates by incorporation of EdU at 12.5 and 16.5dpc. Immunofluorescence for SOX2 (*green*) and EdU-labelling (*red*) after 1 h00 EdU pulse. At 12.5dpc all cells express SOX2 and a high proportion is dividing. At 16.5dpc, progenitors/stem cells are dorsally restricted to cells lining the cleft and proliferation is significantly reduced. (**b**) Quantification of EdU incorporation. At 12.5 dpc, 40.5 % (SD = 7.5 n = 6) of SOX2^{+ve} cells in RP have incorporated EdU, whereas at 16.5 dpc, as SOX9 is up-regulated, only 19.2 % (SD = 2.1, n = 6) do so (*asterisks* in **b**?)

Characterization of AdSC in the Hypothalamo-Pituitary Axis (Fig. 3)

Hypothalamus

Initial investigations of cell proliferation patterns in the post-natal hypothalamus revealed the presence of dividing cells, in particular around the third ventricle, just above the ME. Cell proliferation could be stimulated by infusion of different factors, such as BDNF (Pencea et al. 2001), EFG and FGF (Xu et al. 2005), IGF (Perez-Martin et al. 2010) and CNTF (Kokoeva et al. 2005). In addition, label-retaining experiments suggested active neurogenesis in the hypothalamus (Kokoeva et al. 2005, 2007; Xu et al. 2005; Perez-Martin et al. 2010), defining a

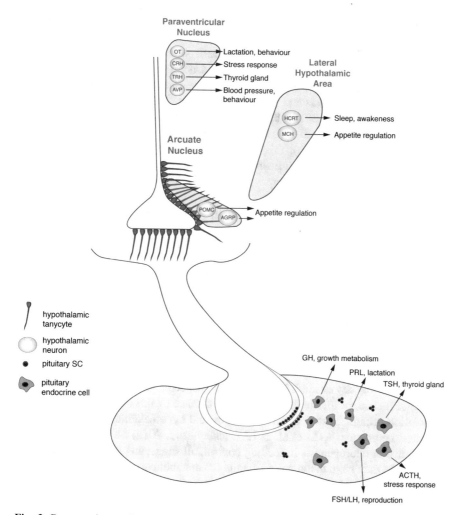

Fig. 3 Regenerative medicine in the hypothalamo-pituitary axis: endogenous stem cells and in vitro differentiated cell types. Endogenous stem cell (SC) populations (*red*) and differentiated progeny (*brown*) obtained in vitro are represented. In the hypothalamus, tanycytes, specialized glial cells located at the base of the third ventricle, form a diet-responsive stem cell population. In the pituitary, stem cells are present in the epithelium lining the cleft and are also scattered in the anterior lobe, sometimes as rosettes. Both hypothalamic neurons and pituitary endocrine cells have been obtained in vitro from ESC and iPSC. In the hypothalamus, transplantation of neurons could be used to modulate feeding behaviour to treat obesity for example, but the range of neurons obtained in vitro could potentially be used to manipulate a range of hypothalamic functions. In the pituitary, transplantation of endocrine cells would represent a significant improvement over existing replacement or substitution therapies

third neurogenic niche in the brain along with the sub-ventricular zone of the lateral ventricles and the dentate gyrus of the hippocampus. Kokoeva et al. (2005) were the first to suggest that hypothalamic neurogenesis was physiologically relevant, as they demonstrated a role for newly generated neurons in feeding control. More recently, lineage tracing experiments have firmly established the existence of active hypothalamic neurogenesis and gliogenesis (Lee et al. 2012; Li et al. 2012; Haan et al. 2013; Robins et al. 2013a, b; see also Blackshaw 2016) Precise dissections of the third ventricle sub-ventricular zone revealed that α-tanycytes are hypothalamic AdSC (Robins et al. 2013b); this area is also where cell proliferation is most efficiently stimulated by IGF infusion (Perez-Martin et al. 2010). WNT signalling appears to regulate tanycyte generation post-natally (Wang et al. 2012). In addition, it has been suggested that some stem cells may reside in the parenchyma (Robins et al. 2013a). Following the initial report by Kokoeva et al., investigations have mostly focused on the role of hypothalamic neurogenesis in feeding control (Lee et al. 2012; Li et al. 2012; McNay et al. 2012; Haan et al. 2013). Stem cells are responsive to diet and a high fat diet reproducibly impairs neurogenesis in the arcuate nucleus (Li et al. 2012; McNay et al. 2012; Lee et al. 2014). Further results strengthen the association between pathological weight gain and neurogenesis impairment. Leptin is a satiety hormone, secreted by adipose cells and transported to the hypothalamus where it participates in appetite regulation. Obesity is associated with leptin insensitivity and, interestingly, leptin deficiency in mice results in impaired neurogenesis (McNay et al. 2012). Moreover, it has been known for some time that obesity is associated with hypothalamic inflammation. As this inflammation precedes obesity onset, it is increasingly suspected to be the cause, rather than the consequence, of diet-induced metabolic disease (Valdearcos et al. 2015). The association of high fat diet, leptin deficiency and hypothalamic inflammation with impaired neurogenesis (Li et al. 2012; McNay et al. 2012) further highlights the importance of neurogenesis in feeding control, all suggesting that manipulation of this process may have therapeutic benefits for metabolic syndromes.

In addition to feeding control, studies in seasonal mammals support a role for hypothalamic neurogenesis in the control of reproduction (Batailler et al. 2015; Ebling 2015). Moreover, there is now evidence that systemic aging is initiated by hypothalamic inflammation (Zhang et al. 2013). The relevant targets of this inflammation, initiated by microglia, are the GnRH neurons, causing them to secrete less GnRH. The authors show that this decrease in GnRH contributes to systemic aging and that it is associated with decreased neurogenesis in both the hypothalamus and hippocampus. Its contribution to these processes is so far unclear, but GnRH administration rescues both GnRH levels and neurogenesis, demonstrating at least a correlation (Zhang et al. 2013). It would now be of interest to investigate the contribution of hypothalamic SC in other life-changing, physiological contexts, such as puberty and pregnancy, where the organism needs to adapt to and trigger, in the case of puberty, a new physiological status.

Pituitary

In the adult gland, the persistence of an epithelial cell layer lining the pituitary cleft (the remnant of the embryonic RP epithelium surrounding the lumen), and the maintenance of SOX2 and SOX9 expression in this epithelial cell layer, was a good argument for the persistence of a progenitor population (Fauquier et al. 2008). The capacity of these cells to form spheres or colonies in vitro, an assay used to characterize progenitors in different tissues further reinforced this hypothesis (Chen et al. 2005; Lepore et al. 2005; Fauquier et al. 2008). Recently, lineage tracing analysis using either $Sox2$ or $Sox9^{CreERT2}$ targeted alleles definitively demonstrated their presence in the adult (Andoniadou et al. 2013; Rizzoti et al. 2013; see also Vankelecom 2016).

Under normal physiological conditions, adult pituitary stem cells proliferate and differentiate very little, suggesting that most cell turnover is due to endocrine cell division (Fauquier et al. 2008; Andoniadou et al. 2013; Rizzoti et al. 2013). Induction of apoptosis upon cell division, using specific genetic tools, confirmed this hypothesis by demonstrating that corticotroph turnover relies on the proliferation of differentiated cells (Langlais et al. 2013). A physiological role for pituitary stem cells was initially suggested by studies investigating models of pituitary target organ ablation (Nolan and Levy 2006). It has been known for some time that ablation of the adrenals and/or gonads triggered a transient mitotic wave in the gland, followed by generation of increased numbers of endocrine cells; moreover, these were specifically the type that normally regulates the ablated organ. While differentiated endocrine cells can divide, but do so rarely, Nolan and Levy were the first to observe that, after adrenalectomy and/or gonadectomy, proliferation is essentially restricted to non-endocrine cells of an immature appearance. Moreover, double ablations do not have an additive proliferative effect, suggesting that a single population of undifferentiated cells responds to both adrenalectomy and gonadectomy (Nolan and Levy 2006). More recently, diphtheria toxin-mediated endocrine cell ablation experiments confirmed these results by showing a mobilization of SOX2-positive cells, with a transient induction of proliferation and, presumably, differentiation. The extent of endocrine cell regeneration was limited and initial endocrine cell numbers never fully recovered; however, this could indicate that a partial recovery is physiologically sufficient (Fu et al. 2012; Fu and Vankelecom 2012). Finally, lineage-tracing experiments performed after pituitary target organ ablation firmly demonstrated that AdSCs both proliferate and differentiate, exclusively to give rise to the required cell type, such as corticotrophs (Rizzoti et al. 2013). This finding definitively established the regenerative potential of pituitary stem cells but also showed that they seem to be mostly, and perhaps only, mobilized under physiological challenge. Dissection of the mechanisms underlying mobilization is now required to investigate whether it would be possible to directly stimulate stem cells for therapeutic purposes. Further investigation into different, more physiological "challenging" situations such as, for example, pregnancy and lactation will also be necessary to fully explore potential roles for AdSCs.

While stem cells appear quiescent under normal physiological situations, their proliferation or at least activation may also be involved in tumorigenesis. The WNT signalling pathway is an important regulator of embryonic development but also of many AdSC populations (see Andoniadou 2016). Moreover, deregulation of the pathway is associated with cancer formation, a notable example being colorectal tumors (Clevers and Nusse 2012). In mice, it had been shown that expression of a degradation-resistant form of βcatenin, an important transducer of WNT signalling, in RP induced the formation of tumors resembling the mostly pediatric pituitary craniopharyngiomas (Gaston-Massuet et al. 2011). Postnatal induction of this constitutively active form of βcatenin in SOX2-positive cells also resulted in tumor formation, demonstrating the tumor-forming potential of the AdSC compartment. Tumors were, however, not composed of mutant stem cells, which instead induced neighboring cells to form them by a paracrine mechanism (Andoniadou et al. 2013). In resected human pituitary adenomas, a side population could be isolated. SOX2 expression was upregulated and spheres could be formed, suggesting the presence of a SOX2-positive progenitor population in tumors (Mertens et al. 2015). However, the role of SOX2-positive cells in adenoma formation is at present unclear.

In vitro Recapitulation of Ontogenesis in the Hypothalamo-Pituitary Axis (Fig. 3)

Two different methods have mostly been used to direct ESC differentiation towards differentiated cell types: tri-dimensional floating aggregates with relatively minimal exogenous treatments, where cell interactions underlie self-patterning into a defined embryonic structure with spectacular results, such as formation of an optic cup realized in the lab of the late Yoshiki Sasai (Eiraku et al. 2011); and two-dimensional cultures where sequential exogenous treatments guide cells through subsequent embryonic fates. Both methods have been successfully used to reproduce the embryonic events described above and obtain both human and murine neuroendocrine and endocrine cells.

Induction of Hypothalamic Identity from 3D ESC Aggregates

Mouse ESC aggregates were initially assayed for hypothalamic neuron generation (Ohyama et al. 2005). Neural progenitor fate was first induced under serum-free conditions (Okabe et al. 1996), followed by SHH and BMP7 treatment that resulted in the generation of ventral hypothalamic neurons (Ohyama et al. 2005). In a later study, culture under strictly defined chemical conditions to minimize exogenous signals was shown to be key to obtain rostral-most hypothalamic character in mouse ESC aggregates, which adopt an embryonic neuroepithelium-like morphology because neural anterior fate is acquired by default in such conditions (Wataya et al. 2008). Removal of insulin was crucial as it was shown to activate the Akt pathway, which would otherwise inhibit hypothalamic induction. Further treatment with SHH induces rostro-ventral hypothalamic identity, with a strong expression of RAX. Selection and re-aggregation of RAX positive cells, without SHH addition, resulted in the differentiation of morphologically mature, secreting vasopressin neurons. Moreover, further treatment of the re-aggregated cells with SHH induces efficient differentiation of other types of hypothalamic neurons (Wataya et al. 2008; Merkle et al. 2015). Very recently, human ESC and iPSC were used to generate hypothalamic neurons in aggregates (Merkle et al. 2015; Wang et al. 2015), following the previously dribed protocol (Wataya et al. 2008). Hypothalamic neurons were successfully obtained, but important differences were noted. These included the requirement for insulin for initial aggregate survival, along with an Akt inhibitor (Merkle et al. 2015), whereas Wang et al. (2015) used SMAD inhibitors to inactivate BMP and TGβ/Nodal/Activin pathways to promote neural differentiation. The major types of hypothalamic neurons were obtained, and further culturing with supporting mouse glia showed that these adopted morphologies similar to their in vivo counterpart, suggesting a mature phenotype. However there is a large variability in induction efficiency between experiments (Merkle et al. 2015).

Generation of Hypothalamic Neurons from 2D ESC Cultures

Both Wang et al. (2015) and Merkle et al. (2015) subsequently developed 2D differentiation protocols to reduce culture duration, providing an easier substrate for expansion than individual aggregates and also reducing variability in differentiation rate. Human ESC and iPSC were induced toward a hypothalamic progenitor fate by sequentially performing dual SMAD inhibition while activating ventralizing SHH signalling (Merkle et al. 2015; Wang et al. 2015). Subsequently, posteriorizing WNT signals were blocked (Merkle et al. 2015) or the NOTCH pathway was inhibited (Wang et al. 2015) to induce arcuate-ventral fate. Further culture on cortical mouse glia supported neuronal differentiation and led to the generation of the major types of hypothalamic neurons. These showed both the expected morphology and neuropeptide expression (Merkle et al. 2015), whereas addition of BDNF induced differentiation of progenitors into functional neurons typical of the arcuate nucleus. This process was further enhanced by co-culture with mouse astrocytes (Wang et al. 2015). Neurons were then transplanted into newborn mouse brains. The human MCH and orexin cells were maintained in the adult where they displayed features resembling synapses with mouse neurons, suggesting functional integration (Merkle et al. 2015).

Induction of RP from 3D Aggregates

Following their success with generation of hypothalamic-like tissue (Wataya et al. 2008), Suga et al. (2011) reasoned that, by further inducing rostral character, they should obtain a domain equivalent to the anterior neural ridge apposed to a neurectoderm-like tissue, where RP formation may be induced (Ochiai et al. 2015; see Suga 2016). Large ESC aggregates were made in growth factor-free medium in the presence of SHH agonist, resulting in formation of a superficial ectodermal layer surrounding a RAX-positive neurectodermal domain. Subsequently, ecto-dermal patches thickened, RP marker expression was induced, and invagination of RP-like structures was observed, more efficiently in the presence of both BMP and FGF (Ochiai et al. 2015). The positive action of these two factors fits with in vivo requirements, because these are secreted from the infundibulum and neces-sary for RP induction and maintenance, respectively (Ericson et al. 1998; Treier et al. 1998). Inhibition of NOTCH signalling resulted, as reported in vivo (Zhu et al. 2006; Kita et al. 2007), in efficient differentiation of corticotrophs, one of the earliest cell types to differentiate. Transplantation of differentiated aggregates under the kidney capsule in hypophysectomised mice improved cortisol deficiency symptoms, which demonstrated the functionality of the ESC-derived corticotrophs (Suga et al. 2011). Obtaining other endocrine cell types was initially less efficient (Suga et al. 2011); however, addition of BMP and FGF improves both corticotroph and Pit-1 lineage endocrine cell differentiation (Ochiai et al. 2015).

Induction of placodal identity and differentiation of endocrine cells from 2D cultures

Human ESC were used to devise a sequential protocol directing cells toward a placodal fate (Dincer et al. 2013; see Studer 2016). Dual SMAD inhibition was initially performed to impart neural fate, and subsequent de-repression in combi-nation with FGF inhibition was sufficient for acquisition of placodal identity. Modulation of FGF, BMP and SHH pathways further allowed specification of lens, trigeminal and pituitary placode identity. As shown earlier (Suga et al. 2011), treatment with an HH agonist induced pituitary placode fate and subsequently an RP gene expression profile. Corticotroph differentiation followed, whereas differentiation of somatotrophs and gonadotrophs relied on NOTCH path-way inhibition. In vivo secretion was demonstrated after sub-cutaneous transplant-ation in mice (Dincer et al. 2013).

Perspectives

Differentiation Strategies

It takes a minimum of 18 days to observe ACTH-positive cells from ESC aggre-gates, and maturation of hypothalamic neurons can take much longer (Merkle et al. 2015). Moreover, the progression through different developmental stages implies some degree of heterogeneity in the culture. Finally, and probably in consequence, the percentage of desired cells obtained is often low. For disease modelling, drug screening and clinical use of these variables should be improved. Culture conditions can be modified and/or the starting cell type, as we will discuss here.

It has already been observed that co-culture with supporting cells, such as cortical glia (Merkle et al. 2015) and specifically astrocytes (Wang et al. 2015), improves differentiation rates of hypothalamic neurons. It has been proposed that co-culture with neurons may also help (Merkle et al. 2015). Differentiation of pituitary endocrine cells may similarly be improved by co-culture with folliculo-stellate cells, a heterogeneous population comprising stem cells but also endocrine-supporting cells (Allaerts and Vankelecom 2005) and/or endocrine cells. Different endocrine cell lines exist and one could examine whether co-culture with these might favor differentiation toward each particular cell type. In addition, generation of 3D pituitary mixed cell type-aggregates has been described in which tissue-like organization and endocrine secretions seem to mimic the in vivo situation; moreover, these characteristics are efficiently maintained over several weeks (Denef et al. 1989). Co-culture in these conditions may help progenitors to differentiate more efficiently.

In aggregates, the absence of vascularization may compromise endocrine differentiation and maturation, as observed in organoids, and affect cell survival more generally in the expanding structures (Lancaster and Knoblich 2014). There are, of course, many other structural and cellular components of the endogenous stem cell niche that are missing in these aggregates. Therefore, reconstitution of the stem cells' microenvironment in vitro has been the focus of several investigations. Biomaterials have been developed to support co-culture to supply vascularization, for example, but also to allow applications of factors, either extracellular matrix or signalling molecules. Recent and promising droplet-based microfluidic strategies have been described that can be used in either 3D or 2D cultures (Allazetta and Lutolf 2015). These in vitro micro-niches are of interest for drug screening because they are scalable, but they may also offer a suitable substrate to obtain cells for transplantation. Perhaps these would further allow organization into a pattern typical of the anterior pituitary.

Starting from pluripotent stem cells, either ESC or iPSC implies a long developmental "journey" to reach a mature terminally differentiated state. Since populations of progenitors have been characterized in both compartments of the hypothalamo-pituitary axis, a "short cut" would be to start from somatic cells that are then directly re-programmed into hypothalamic or pituitary specific stem cells. Similar protocols were initially described for NSCs (Kim et al. 2011; Ring et al. 2012; Thier et al. 2012) and many other cell types since. Self-renewal will be comparable in ESC and iPSC, but the differentiation potential is now limited to the cells of interest. This implies characterization of the factors necessary and sufficient to impart the desired identity upon re-programming.

Implantation in Homotypic Locations

Once the cell types of interest have been efficiently generated in vitro, they can be transplanted back to restore function. Up to now, transplantations of either pituitary or hypothalamic ES-derived cells have been realised in heterotopic locations; under the kidney capsule (Suga et al. 2011), or subcutaneously (Dincer et al. 2013) for endocrine cells, and in the lateral ventricle or brain parenchyma for hypothalamic neurons (Merkle et al. 2015).

While Suga et al. (2011) successfully improved some hypophysectomy symptoms by implanting cells under the kidney capsule, the pituitary and hypothalamus are physically and functionally connected: proper regulation of endocrine secretions requires connection with the hypothalamus. In pilot experiments in the 1950s, Harris and collaborators indeed demonstrated that anterior pituitary transplantation away from its normal location, such as under the kidney capsule, resulted in chronic PRL secretion and essentially loss of secretion of the other hormones. In contrast, grafts implanted near the pituitary stalk, shortly after hypophysiectomy, resulted in regeneration of the portal system and this was associated with functional integration of the graft (Harris and Jacobsohn 1951). Therefore, transphenoidal endoscopic cell transplantation in the human pituitary close to the pituitary stalk should promote adequate integration and control (see Studer and Tabar 2016).

Hypothalamic neurons need to be able to make relevant connections, particularly those controlling the pituitary via the ME. Transplantations of cells or grafts have been realized in the hypothalamus with spectacular results. Placement of pre-optic area (POA) grafts containing GnRH neurons close to the ME successfully restores reproductive function in the *gnrh1* mutant hypogonadal mice, independently of the sex of the donor. As expected, success appears to rely on accession of GnRH axons to the ME (Gibson et al. 1984; Charlton 2004). Anterior hypothalamic implants comprising the suprachiasmatic nucleus restore periodicity in animals rendered arrhythmic by hypothalamic lesions (Sollars et al. 1995). More recently, immature hypothalamic neurons and progenitors were transplanted into the hypothalamic parenchyma of early postnatal brains of leptin receptor-deficient mice. These resulted in functional integration and partial restoration of leptin responsiveness in the adult (Czupryn et al. 2011). However, transplantations close to the ME might offer better results, because arcuate nucleus grafts transplanted into the third ventricle, close to the ME, are associated with comparatively better anti-obesity effects in obese rats (Ono et al. 1990; Fetissov et al. 2000). Integration near a site where the blood-brain barrier is interrupted might allow access to peripheral signals, such as leptin in this context and, therefore, better functionality. All these studies show that hypothalamic implantation can restore function; therefore, they offer hope that stem cell-derived hypothalamic neurons transplanted near the ME would be successful. However, as demonstrated in mice, damage to the ME/stalk, such as that observed after traumatic brain injuries, is probably causative of pituitary deficiencies (Osterstock et al. 2014). Therefore, just as transplantation close to this site has been proposed to improve the functionality of both pituitary and hypothalamic transplants, care should be taken not to damage this fragile structure.

Conclusion

The generation of endocrine cells from ESC and their subsequent transplantation in vivo represent exciting progress towards the use of regenerative medicine to treat endocrine deficits or manipulate endocrine outputs. Improvement of differentiation efficiencies and transplantation in homotypic locations should in the near future demonstrate whether regenerative therapies are suitable for clinical use to treat neuro-endocrinological disorders.

References

Alatzoglou KS, Webb EA, Le Tissier P, Dattani MT (2014) Isolated growth hormone deficiency (GHD) in childhood and adolescence: recent advances. Endocr Rev 35:376–432

Allaerts W, Vankelecom H (2005) History and perspectives of pituitary folliculo-stellate cell research. Eur J Endocrinol 153:1–12

Allazetta S, Lutolf MP (2015) Stem cell niche engineering through droplet microfluidics. Curr Opin Biotechnol 35:86–93

Andoniadou CL (2016) Pituitary stem cells during normal physiology and disease. In: Pfaff D, Christen Y (eds) Stem cells in neuroendocrinology. Springer, Heidelberg

Andoniadou CL, Matsushima D, Mousavy Gharavy SN, Signore M, Mackintosh AI, Schaeffer M, Gaston-Massuet C, Mollard P, Jacques TS, Le Tissier P, Dattani MT, Pevny LH, Martinez-Barbera JP (2013) Sox2(+) stem/progenitor cells in the adult mouse pituitary support organ homeostasis and have tumor-inducing potential. Cell Stem Cell 13:433–445

Batailler M, Derouet L, Butruille L, Migaud M (2015) Sensitivity to the photoperiod and potential migratory features of neuroblasts in the adult sheep hypothalamus. Brain Struct Funct doi:10.1007/ss00429-015-1101-0

Bedont JL, Newman EA, Blackshaw S (2015) Patterning, specification, and differentiation in the developing hypothalamus. Wiley interdisciplinary reviews. Dev Biol 4:445–468

Blackshaw S (2016) Regulation of body weight and metabolism by tanycyte-derived neurogenesis in young adult mice. In: Pfaff D, Christen Y (eds) Stem cells in neuroendocrinology. Springer, Heidelberg

Bolborea M, Dale N (2013) Hypothalamic tanycytes: potential roles in the control of feeding and energy balance. Trends Neurosci 36:91–100

Castinetti F, Davis SW, Brue T, Camper SA (2011) Pituitary stem cell update and potential implications for treating hypopituitarism. Endocr Rev 32:453–471

Charlton H (2004) Neural transplantation in hypogonadal (hpg) mice—physiology and neurobiology. Reproduction 127:3–12

Chen J, Hersmus N, Van Duppen V, Caesens P, Denef C, Vankelecom H (2005) The adult pituitary contains a cell population displaying stem/progenitor cell and early embryonic characteristics. Endocrinology 146:3985–3998

Clevers H, Nusse R (2012) Wnt/beta-catenin signaling and disease. Cell 149:1192–1205

Czupryn A, Zhou YD, Chen X, McNay D, Anderson MP, Flier JS, Macklis JD (2011) Transplanted hypothalamic neurons restore leptin signaling and ameliorate obesity in db/db mice. Science 334:1133–1137

Davis SW, Mortensen AH, Camper SA (2011) Birthdating studies reshape models for pituitary gland cell specification. Dev Biol 352:215–227

Denef C, Maertens P, Allaerts W, Mignon A, Robberecht W, Swennen L, Carmeliet P (1989) Cell-to-cell communication in peptide target cells of anterior pituitary. Methods Enzymol 168:47–71

Dincer Z, Piao J, Niu L, Ganat Y, Kriks S, Zimmer B, Shi SH, Tabar V, Studer L (2013) Specification of functional cranial placode derivatives from human pluripotent stem cells. Cell Rep 5:1387–1402

Drouin J (2016) Epigenetic mechanisms of pituitary cell fate specification. In: Pfaff D, Christen Y (eds) Stem cells in neuroendocrinology. Springer, Heidelberg

Ebling FJ (2015) Hypothalamic control of seasonal changes in food intake and body weight. Front Neuroendocrinol 37:97–107

Eiraku M, Takata N, Ishibashi H, Kawada M, Sakakura E, Okuda S, Sekiguchi K, Adachi T, Sasai Y (2011) Self-organizing optic-cup morphogenesis in three-dimensional culture. Nature 472:51–56

Ericson J, Norlin S, Jessell TM, Edlund T (1998) Integrated FGF and BMP signaling controls the progression of progenitor cell differentiation and the emergence of pattern in the embryonic anterior pituitary. Development 125:1005–1015

Fauquier T, Rizzoti K, Dattani M, Lovell-Badge R, Robinson IC (2008) SOX2-expressing progenitor cells generate all of the major cell types in the adult mouse pituitary gland. Proc Natl Acad Sci USA 105:2907–2912

Fetissov SO, Gerozissis K, Orosco M, Nicolaidis S (2000) Synergistic effect of arcuate and raphe nuclei graft to alleviate insulinemia and obesity in Zucker rats. Acta Diabetol 37:65–70

Fox IJ, Daley GQ, Goldman SA, Huard J, Kamp TJ, Trucco M (2014) Stem cell therapy. Use of differentiated pluripotent stem cells as replacement therapy for treating disease. Science 345:1247391

French A et al (2015) Enabling consistency in pluripotent stem cell-derived products for research and development and clinical applications through material standards. Stem Cells Translat Med 4:217–223

Fu Q, Vankelecom H (2012) Regenerative capacity of the adult pituitary: multiple mechanisms of lactotrope restoration after transgenic ablation. Stem Cells Dev 21:3245–3257

Fu Q, Gremeaux L, Luque RM, Liekens D, Chen J, Buch T, Waisman A, Kineman R, Vankelecom H (2012) The adult pituitary shows stem/progenitor cell activation in response to injury and is capable of regeneration. Endocrinology 153:3224–3235

Gaston-Massuet C, Andoniadou CL, Signore M, Jayakody SA, Charolidi N, Kyeyune R, Vernay B, Jacques TS, Taketo MM, Le Tissier P, Dattani MT, Martinez-Barbera JP (2011) Increased Wingless (Wnt) signaling in pituitary progenitor/stem cells gives rise to pituitary tumors in mice and humans. Proc Natl Acad Sci USA 108:11482–11487

Gibson MJ, Krieger DT, Charlton HM, Zimmerman EA, Silverman AJ, Perlow MJ (1984) Mating and pregnancy can occur in genetically hypogonadal mice with preoptic area brain grafts. Science 225:949–951

Goto M, Hojo M, Ando M, Kita A, Kitagawa M, Ohtsuka T, Kageyama R, Miyamoto S (2015) Hes1 and Hes5 are required for differentiation of pituicytes and formation of the neuro-hypophysis in pituitary development. Brain Res 1625:206–217

Haan N, Goodman T, Najdi-Samiei A, Stratford CM, Rice R, El Agha E, Bellusci S, Hajihosseini MK (2013) Fgf10-expressing tanycytes add new neurons to the appetite/energy-balance regulating centers of the postnatal and adult hypothalamus. J Neurosci 33:6170–6180

Hannon MJ, Crowley RK, Behan LA, O'Sullivan EP, O'Brien MM, Sherlock M, Rawluk D, O'Dwyer R, Tormey W, Thompson CJ (2013) Acute glucocorticoid deficiency and diabetes insipidus are common after acute traumatic brain injury and predict mortality. J Clin Endocrinol Metab 98:3229–3237

Harris GW, Jacobsohn D (1951) Functional grafts of the anterior pituitary gland. J Physiol 113: 35p–36p

Himes AD, Raetzman LT (2009) Premature differentiation and aberrant movement of pituitary cells lacking both Hes1 and Prop1. Dev Biol 325:151–161

Hsu PD, Lander ES, Zhang F (2014) Development and applications of CRISPR-Cas9 for genome engineering. Cell 157:1262–1278

Huch M, Koo BK (2015) Modeling mouse and human development using organoid cultures. Development 142:3113–3125

Japon MA, Rubinstein M, Low MJ (1994) In situ hybridization analysis of anterior pituitary hormone gene expression during fetal mouse development. J Histochem Cytochem 42: 1117–1125

Kamachi Y, Kondoh H (2013) Sox proteins: regulators of cell fate specification and differentiation. Development 140:4129–4144

Kelberman D, Rizzoti K, Lovell-Badge R, Robinson IC, Dattani MT (2009) Genetic regulation of pituitary gland development in human and mouse. Endocr Rev 30:790–829

Khonsari RH et al (2013) The buccohypophyseal canal is an ancestral vertebrate trait maintained by modulation in sonic hedgehog signaling. BMC Biol 11:27

Kim J, Efe JA, Zhu S, Talantova M, Yuan X, Wang S, Lipton SA, Zhang K, Ding S (2011) Direct reprogramming of mouse fibroblasts to neural progenitors. Proc Natl Acad Sci USA 108:7838–7843

Kita A, Imayoshi I, Hojo M, Kitagawa M, Kokubu H, Ohsawa R, Ohtsuka T, Kageyama R, Hashimoto N (2007) Hes1 and Hes5 control the progenitor pool, intermediate lobe specification, and posterior lobe formation in the pituitary development. Mol Endocrinol 21:1458–1466

Kokoeva MV, Yin H, Flier JS (2005) Neurogenesis in the hypothalamus of adult mice: potential role in energy balance. Science 310:679–683

Kokoeva MV, Yin H, Flier JS (2007) Evidence for constitutive neural cell proliferation in the adult murine hypothalamus. J Comp Neurol 505:209–220

Konadhode RR, Pelluru D, Shiromani PJ (2014) Neurons containing orexin or melanin concentrating hormone reciprocally regulate wake and sleep. Front Syst Neuro 8:244

Lancaster MA, Knoblich JA (2014) Organogenesis in a dish: modeling development and disease using organoid technologies. Science 345:1247125

Langlais D, Couture C, Kmita M, Drouin J (2013) Adult pituitary cell maintenance: lineage-specific contribution of self-duplication. Mol Endocrinol 27:1103–1112

Lee DA et al (2014) Dietary and sex-specific factors regulate hypothalamic neurogenesis in young adult mice. Front Neurosci 8:157

Lee DA, Bedont JL, Pak T, Wang H, Song J, Miranda-Angulo A, Takiar V, Charubhumi V, Balordi F, Takebayashi H, Aja S, Ford E, Fishell G, Blackshaw S (2012) Tanycytes of the hypothalamic median eminence form a diet-responsive neurogenic niche. Nat Neurosci 15: 700–702

Lepore DA, Roeszler K, Wagner J, Ross SA, Bauer K, Thomas PQ (2005) Identification and enrichment of colony-forming cells from the adult murine pituitary. Exp Cell Res 308:166–176

Levy A (2002) Physiological implications of pituitary trophic activity. J Endocrinol 174:147–155

Li J, Tang Y, Cai D (2012) IKKbeta/NF-kappaB disrupts adult hypothalamic neural stem cells to mediate a neurodegenerative mechanism of dietary obesity and pre-diabetes. Nat Cell Biol 14: 999–1012

Lu F, Kar D, Gruenig N, Zhang ZW, Cousins N, Rodgers HM, Swindell EC, Jamrich M, Schuurmans C, Mathers PH, Kurrasch DM (2013) Rax is a selector gene for mediobasal hypothalamic cell types. J Neurosci 33:259–272

McNay DE, Briancon N, Kokoeva MV, Maratos-Flier E, Flier JS (2012) Remodeling of the arcuate nucleus energy-balance circuit is inhibited in obese mice. J Clin Invest 122:142–152

McShane SG, Mole MA, Savery D, Greene ND, Tam PP, Copp AJ (2015) Cellular basis of neuroepithelial bending during mouse spinal neural tube closure. Dev Biol 404:113–124

Merkle FT, Maroof A, Wataya T, Sasai Y, Studer L, Eggan K, Schier AF (2015) Generation of neuropeptidergic hypothalamic neurons from human pluripotent stem cells. Development 142: 633–643

Mertens FM, Gremeaux L, Chen J, Fu Q, Willems C, Roose H, Govaere O, Roskams T, Cristina C, Becu-Villalobos D, Jorissen M, Vander Poorten V, Bex M, van Loon J, Vankelecom H (2015) Pituitary tumors contain a side population with tumor stem cell-associated characteristics. Endocr Relat Cancer 22:481–504

Mollard P, Hodson DJ, Lafont C, Rizzoti K, Drouin J (2012) A tridimensional view of pituitary development and function. Trends Endocrinol Metab 23:261–269

Nolan LA, Levy A (2006) A population of non-luteinising hormone/non-adrenocorticotrophic hormone-positive cells in the male rat anterior pituitary responds mitotically to both gonadectomy and adrenalectomy. J Neuroendocrinol 18:655–661

Ochiai H, Suga H, Yamada T, Sakakibara M, Kasai T, Ozone C, Ogawa K, Goto M, Banno R, Tsunekawa S, Sugimura Y, Arima H, Oiso Y (2015) BMP4 and FGF strongly induce differentiation of mouse ES cells into oral ectoderm. Stem Cell Res 15:290–298

Ohyama K, Ellis P, Kimura S, Placzek M (2005) Directed differentiation of neural cells to hypothalamic dopaminergic neurons. Development 132:5185–5197

Okabe S, Forsberg-Nilsson K, Spiro AC, Segal M, McKay RD (1996) Development of neuronal precursor cells and functional postmitotic neurons from embryonic stem cells in vitro. Mech Dev 59:89–102

Ono K, Kawamura K, Shimizu N, Ito C, Plata-Salaman CR, Ogawa N, Oomura Y (1990) Fetal hypothalamic brain grafts to the ventromedial hypothalamic obese rats: an immunohistochemical, electrophysiological and behavioral study. Brain Res Bull 24:89–96

Osterstock G, El Yandouzi T, Romano N, Carmignac D, Langlet F, Coutry N, Guillou A, Schaeffer M, Chauvet N, Vanacker C, Galibert E, Dehouck B, Robinson IC, Prevot V, Mollard P, Plesnila N, Mery PF (2014) Sustained alterations of hypothalamic tanycytes during posttraumatic hypopituitarism in male mice. Endocrinology 155:1887–1898

Pearson CA, Placzek M (2013) Development of the medial hypothalamus: forming a functional hypothalamic-neurohypophyseal interface. Curr Top Dev Biol 106:49–88

Pearson CA, Ohyama K, Manning L, Aghamohammadzadeh S, Sang H, Placzek M (2011) FGF-dependent midline-derived progenitor cells in hypothalamic infundibular development. Development 138:2613–2624

Pencea V, Bingaman KD, Wiegand SJ, Luskin MB (2001) Infusion of brain-derived neurotrophic factor into the lateral ventricle of the adult rat leads to new neurons in the parenchyma of the striatum, septum, thalamus, and hypothalamus. J Neurosci 21:6706–6717

Perez-Martin M, Cifuentes M, Grondona JM, Lopez-Avalos MD, Gomez-Pinedo U, Garcia-Verdugo JM, Fernandez-Llebrez P (2010) IGF-I stimulates neurogenesis in the hypothalamus of adult rats. Eur J Neurosci 31:1533–1548

Prevot V, Bellefontaine N, Baroncini M, Sharif A, Hanchate NK, Parkash J, Campagne C, de Seranno S (2010) Gonadotrophin-releasing hormone nerve terminals, tanycytes and neurohaemal junction remodelling in the adult median eminence: functional consequences for reproduction and dynamic role of vascular endothelial cells. J Neuroendocrinol 22:639–649

Ring KL, Tong LM, Balestra ME, Javier R, Andrews-Zwilling Y, Li G, Walker D, Zhang WR, Kreitzer AC, Huang Y (2012) Direct reprogramming of mouse and human fibroblasts into multipotent neural stem cells with a single factor. Cell Stem Cell 11:100–109

Rizzoti K (2015) Genetic regulation of murine pituitary development. J Mol Endocrinol 54: R55–R73

Rizzoti K, Brunelli S, Carmignac D, Thomas PQ, Robinson IC, Lovell-Badge R (2004) SOX3 is required during the formation of the hypothalamo-pituitary axis. Nat Genet 36:247–255

Rizzoti K, Akiyama H, Lovell-Badge R (2013) Mobilized adult pituitary stem cells contribute to endocrine regeneration in response to physiological demand. Cell Stem Cell 13:419–432

Robins SC, Trudel E, Rotondi O, Liu X, Djogo T, Kryzskaya D, Bourque CW, Kokoeva MV (2013a) Evidence for NG2-glia derived, adult-born functional neurons in the hypothalamus. PLoS One 8:e78236

Robins SC, Stewart I, McNay DE, Taylor V, Giachino C, Goetz M, Ninkovic J, Briancon N, Maratos-Flier E, Flier JS, Kokoeva MV, Placzek M (2013b) alpha-Tanycytes of the adult hypothalamic third ventricle include distinct populations of FGF-responsive neural progenitors. Nat Commun 4:2049

Sadagurski M, Landeryou T, Cady G, Kopchick JJ, List EO, Berryman DE, Bartke A, Miller RA (2015) Growth hormone modulates hypothalamic inflammation in long-lived pituitary dwarf mice. Aging Cell 14:1045–1054

Salvatierra J, Lee DA, Zibetti C, Duran-Moreno M, Yoo S, Newman EA, Wang H, Bedont JL, de Melo J, Miranda-Angulo AL, Gil-Perotin S, Garcia-Verdugo JM, Blackshaw S (2014) The LIM homeodomain factor Lhx2 is required for hypothalamic tanycyte specification and differentiation. J Neurosci 34:16809–16820

Sarkar A, Hochedlinger K (2013) The sox family of transcription factors: versatile regulators of stem and progenitor cell fate. Cell Stem Cell 12:15–30

Schlosser G, Patthey C, Shimeld SM (2014) The evolutionary history of vertebrate cranial placodes II. Evolution of ectodermal patterning. Dev Biol 389:98–119

Schneeberger M, Gomis R, Claret M (2014) Hypothalamic and brainstem neuronal circuits controlling homeostatic energy balance. J Endocrinol 220:T25–T46

Schwartz SD, Regillo CD, Lam BL, Eliott D, Rosenfeld PJ, Gregori NZ, Hubschman JP, Davis JL, Heilwell G, Spirn M, Maguire J, Gay R, Bateman J, Ostrick RM, Morris D, Vincent M, Anglade E, Del Priore LV, Lanza R (2015) Human embryonic stem cell-derived retinal pigment epithelium in patients with age-related macular degeneration and Stargardt's macular dystrophy: follow-up of two open-label phase 1/2 studies. Lancet 385:509–516

Scott CE, Wynn SL, Sesay A, Cruz C, Cheung M, Gomez Gaviro MV, Booth S, Gao B, Cheah KS, Lovell-Badge R, Briscoe J (2010) SOX9 induces and maintains neural stem cells. Nat Neurosci 13:1181–1189

Shamir ER, Ewald AJ (2014) Three-dimensional organotypic culture: experimental models of mammalian biology and disease. Nat Rev Mol Cell Biol 15:647–664

Sollars PJ, Kimble DP, Pickard GE (1995) Restoration of circadian behavior by anterior hypothalamic heterografts. J Neurosci 15:2109–2122

Soukup V, Horacek I, Cerny R (2013) Development and evolution of the vertebrate primary mouth. J Anat 222:79–99

Steinbeck JA, Studer L (2015) Moving stem cells to the clinic: potential and limitations for brain repair. Neuron 86:187–206

Stevenson EL, Corella KM, Chung WC (2013) Ontogenesis of gonadotropin-releasing hormone neurons: a model for hypothalamic neuroendocrine cell development. Front Endocrinol 4:89

Studer L (2016) Human pluripotent-derived lineages for repairing hypopituitarism. In: Pfaff D, Christen Y (eds) Stem cells in neuroendocrinology. Springer, Heidelberg

Studer L, Tabar V (2016) Human pluripotent-derived lineages for repairing hypopituitarism. In: Pfaff D, Christen Y (eds) Stem cells in neuroendocrinology. Springer, Heidelberg

Suga H (2016) Recapitulating hypothalamus and pituitary development using ES/iPS cells. In: Pfaff D, Christen Y (eds) Stem cells in neuroendocrinology. Springer, Heidelberg

Suga H, Kadoshima T, Minaguchi M, Ohgushi M, Soen M, Nakano T, Takata N, Wataya T, Muguruma K, Miyoshi H, Yonemura S, Oiso Y, Sasai Y (2011) Self-formation of functional adenohypophysis in three-dimensional culture. Nature 480:57–62

Takahashi K, Yamanaka S (2006) Induction of pluripotent stem cells from mouse embryonic and adult fibroblast cultures by defined factors. Cell 126:663–676

Thier M, Worsdorfer P, Lakes YB, Gorris R, Herms S, Opitz T, Seiferling D, Quandel T, Hoffmann P, Nothen MM, Brustle O, Edenhofer F (2012) Direct conversion of fibroblasts into stably expandable neural stem cells. Cell Stem Cell 10:473–479

Treier M, Gleiberman AS, O'Connell SM, Szeto DP, McMahon JA, McMahon AP, Rosenfeld MG (1998) Multistep signaling requirements for pituitary organogenesis in vivo. Genes Dev 12: 1691–1704

Trowe MO, Zhao L, Weiss AC, Christoffels V, Epstein DJ, Kispert A (2013) Inhibition of Sox2-dependent activation of Shh in the ventral diencephalon by Tbx3 is required for formation of the neurohypophysis. Development 140:2299–2309

Valdearcos M, Xu AW, Koliwad SK (2015) Hypothalamic inflammation in the control of metabolic function. Annu Rev Physiol 77:131–160

van de Wetering M et al (2015) Prospective derivation of a living organoid biobank of colorectal cancer patients. Cell 161:933–945

Vankelecom H (2016) Pituitary stem cells: quest for hidden functions. In: Pfaff D, Christen Y (eds) Stem cells in neuroendocrinology. Springer, Heidelberg

Vierbuchen T, Wernig M (2012) Molecular roadblocks for cellular reprogramming. Mol Cell 47: 827–838

Wang X et al (2012) Wnt signaling regulates postembryonic hypothalamic progenitor differentiation. Dev Cell 23:624–636

Wang Y, Martin JF, Bai CB (2010) Direct and indirect requirements of Shh/Gli signaling in early pituitary development. Dev Biol 348:199–209

Wang L, Meece K, Williams DJ, Lo KA, Zimmer M, Heinrich G, Martin Carli J, Leduc CA, Sun L, Zeltser LM, Freeby M, Goland R, Tsang SH, Wardlaw SL, Egli D, Leibel RL (2015) Differentiation of hypothalamic-like neurons from human pluripotent stem cells. J Clin Invest 125:796–808

Wataya T, Ando S, Muguruma K, Ikeda H, Watanabe K, Eiraku M, Kawada M, Takahashi J, Hashimoto N, Sasai Y (2008) Minimization of exogenous signals in ES cell culture induces rostral hypothalamic differentiation. Proc Natl Acad Sci USA 105:11796–11801

Xu Y, Tamamaki N, Noda T, Kimura K, Itokazu Y, Matsumoto N, Dezawa M, Ide C (2005) Neurogenesis in the ependymal layer of the adult rat 3rd ventricle. Exp Neurol 192:251–264

Zhang G, Li J, Purkayastha S, Tang Y, Zhang H, Yin Y, Li B, Liu G, Cai D (2013) Hypothalamic programming of systemic ageing involving IKK-beta, NF-kappaB and GnRH. Nature 497: 211–216

Zhao L, Zevallos SE, Rizzoti K, Jeong Y, Lovell-Badge R, Epstein DJ (2012) Disruption of SoxB1-dependent sonic hedgehog expression in the hypothalamus causes septo-optic dysplasia. Dev Cell 22:585–596

Zhu X, Zhang J, Tollkuhn J, Ohsawa R, Bresnick EH, Guillemot F, Kageyama R, Rosenfeld MG (2006) Sustained Notch signaling in progenitors is required for sequential emergence of distinct cell lineages during organogenesis. Genes Dev 20:2739–2753

Printed in the United States
By Bookmasters